THE FOUNDATIONS OF COGNITIVE SCIENCE

The Foundations of Cognitive Science

Edited by
JOÃO BRANQUINHO

CLARENDON PRESS · OXFORD
2001

OXFORD
UNIVERSITY PRESS

Great Clarendon Street, Oxford OX2 6DP

Oxford University Press is a department of the University of Oxford.
It furthers the University's objective of excellence in research, scholarship,
and education by publishing worldwide in

Oxford New York

Athens Auckland Bangkok Bogotá Buenos Aires
Cape Town Chennai Dar es Salaam Delhi Florence Hong Kong Istanbul
Karachi Kolkata Kuala Lumpur Madrid Melbourne Mexico City Mumbai
Nairobi Paris São Paulo Shanghai Singapore Taipei Tokyo Toronto Warsaw
and associated companies in Berlin Ibadan

Oxford is a registered trade mark of Oxford University Press
in the UK and certain other countries

Published in the United States
by Oxford University Press Inc., New York

British Library Cataloguing in Publication Data

Library of Congress Cataloging in Publication Data
The foundations of cognitive science/edited by João Branquinho
p. cm.
Includes bibliographical references and index.
1. Cognitive science. 1. Branquinho, João.

BF311 .F66 2001   153—dc21   2001016335

ISBN 0-19-823890-8 (hbk.)
ISBN 0–19–823889–4 (pbk.)

1 3 5 7 9 10 8 6 4 2

Typeset by Kolam Information Services Pvt. Ltd.
Pondicherry, India
Printed in Great Britain by
Biddles Ltd, Guildford & King's Lynn

Contents

Contents

Notes on Contributors

Ned Block is Professor of Philosophy and Psychology at New York University. He came to NYU in 1996 from MIT where he was Chair of the Philosophy Program. His fields of specialization are: philosophy of mind, foundations of cognitive science, and metaphysics. He has published important papers in these areas and is the editor of *Readings in Philosophy of Psychology* (1980) and co-editor of *The Nature of Consciousness: Philosophical Debates* (1998). He is currently writing a book on consciousness.

Margaret Boden is Professor of Philosophy and Psychology in the School of Cognitive and Computing Sciences, University of Sussex. Her research interests lie in the relations between AI-Life, theoretical psychology, and the philosophy of mind. She has published numerous articles on these issues and is the author of *Artificial Intelligence and Natural Man* (1977), *Computer Models of Mind* (1988), and *The Creative Mind* (1990), and the editor of *The Philosophy of Artificial Intelligence* (1990) and *The Philosophy of Artificial Life* (1996).

Susan Carey is Professor of Psychology at New York University. She works in developmental and cognitive psychology and also in the foundations of cognitive science. Her major research interests are infant cognition, cognitive development, conceptual change in childhood, and face recognition. She has published many papers on these issues and is the author of *Conceptual Change in Childhood* (1985) and co-editor of *The Epigenesis of Mind: Essays on Biology and Cognition* (1990).

Patricia S. Churchland is Professor of Philosophy at the University of California at San Diego. Her fields of specialization are: philosophy of neuroscience, philosophy of mind, philosophy of science, and environmental ethics. She is co-author of *The Computational Brain* (1992) and (with Paul M. Churchland) *On the Contrary: Critical Essays* (1998) and author of *Neurophilosophy: Toward a Unified Science of the Mind-Brain* (1986), as well as numerous articles.

Paul M. Churchland is Professor of Philosophy at the University of California at San Diego. His research interests lie in the philosophy of science, the philosophy of mind, artificial intelligence and cognitive neurobiology, epistemology, and perception. His principal publications include *The Engine of Reason, The Seat of the Soul: A Philosophical Journey into the Brain* (1995), *A Neurocomputational Perspective: The Nature of Mind and the Structure of Science* (1989), and *Matter and Consciousness* (1984).

Antonio R. Damasio is M. W. Van Allen Professor of Neurology at the University of Iowa. He is interested in the neurobiology of the mind, specifically the understanding of the neural systems which subserve memory, language, emotion, and decision-making. He is author of several important papers in these areas and his book *Descartes's Error: Emotion, Reason, and the Human Brain* (1994) has been translated into seventeen languages. *The Feeling of What Happens* (1999) is his most recent book.

Hanna Damasio is Professor of Neurology at the University of Iowa. Her major research interests are the neuroanatomical basis of cognition, the development of new techniques to investigate brain structure in vivo using magnetic resonance, and the neuroanatomical substrates of language, memory and decision-making using the lesion method and PET. She has published extensively on these issues and is the author of *Human Brain Anatomy in Computerized Images* (1995).

Donald Davidson is Willis S. and Marion Slusser Professor of Philosophy at the University of California at Berkeley. He has published a number of widely discussed essays on the philosophy of language, philosophy of mind, and philosophy of action. He is co-editor of *The Semantics of Natural Language* (1972) and *The Logic of Grammar* (1975), and the author of *Essays on Actions and Events* (1980) and *Inquiries into Truth and Interpretation* (1984).

Daniel C. Dennett is Distinguished Arts and Sciences Professor and Director of the Center for Cognitive Studies at Tufts University. He specializes in the philosophy of mind, philosophy of science, and foundations of cognitive science. He has published a number of influential articles and books; the latter include *Brainstorms* (1980), *The Intentional Stance* (1987), *Darwin's Dangerous Idea* (1995), and *Brainchildren* (1998).

Ilya Farber recently completed a Ph.D. in philosophy and cognitive science at the University of California at San Diego. His dissertation explores the historical process whereby science comes to grips with abstract notions such as life and consciousness. He is currently working on the role of conceptual change in the philosophy of mind and in cognitive neuroscience. He is also interested in questions of reductionism (especially in biology) and in Pragmatist epistemology and metaphysics. Works in progress include a critique of the role of thought experiments in philosophy, and a chapter on consciousness for the *Wiley Comprehensive Handbook of Psychology* (coauthored with William Banks and Patricia Churchland).

James T. Higginbotham is Professor of Theoretical Linguistics at the University of Oxford. He came to Oxford from MIT where he taught at the Department of Linguistics and Philosophy. His fields of specialization are:

philosophy of language, philosophy of mind, theoretical linguistics, philosophical logic, and foundations of cognitive science. He has published numerous papers both in philosophy and linguistics and is currently writing a book on semantics.

Christopher Peacocke is Professor of Philosophy at New York University. He was previously Waynflete Professor of Metaphysical Philosophy in the University of Oxford and held a Leverhulme Personal Research Professorship. He specializes in the philosophy of mind and psychology, the philosophy of language, and metaphysics and epistemology. He has published extensively in these areas and his books include *Sense and Content* (1983), *Thoughts: An Essay on Content* (1986), *A Study of Concepts* (1992), and *Being Known* (1999).

Will Peterman is at the Department of Cognitive Science of the University of California at San Diego. He took a BA in Cognitive Science at Pomona College in 1992. His primary study is the effect of environmental and neurological constraints on the evolution of foraging behavior; secondary fields include evolutionary computation and situated natural language processing.

Zenon W. Pylyshyn is Board of Governors Professor of Cognitive Science at Rutgers University. He is interested in the theoretical analysis of the nature of the human computational system which enables humans to reason and perceive the world. He is editor of *Meaning and Cognitive Structure* (1986) and *The Robot's Dilemma Revisited* (1996), and author of *Computation and Cognition: Toward a Foundation for Cognitive Science* (1984), as well as many papers on the foundations of cognitive science.

John R. Searle is Mills Professor of the Philosophy of Mind and Language at the University of California at Berkeley. His major interests are in philosophy of language, metaphysics, philosophy of mind, and philosophy of action. His books include *Intentionality: An Essay in the Philosophy of Mind* (1983), *Minds, Brains, and Science* (1984), *The Rediscovery of the Mind* (1992), and *The Construction of Social Reality* (1995).

INTRODUCTION

João Branquinho

The present volume brings together thirteen new essays dealing with a wide variety of important topics in the foundations of the fascinating multi-disciplinary field of studies currently known as Cognitive Science. The purpose of this Introduction is to provide the reader with a synoptic view of the territory covered by the book, especially the main issues and problems addressed, and to give an outline of the central contributions made by each chapter to their discussion. It is expected that this will help readers, particularly those less familiar with the area, to be able to discern a background of shared theoretical concerns and general assumptions behind the diversity of topics and approaches displayed by the contributed chapters. Given the controversial nature of most issues in the foundations of cognitive science, it could hardly be expected from a description of the territory that it be theoretically neutral; however, we have tried as much as possible to stay close to a set of methodological claims that are very often seen as consensual.

The multi-disciplinary character of the area is clearly reflected in the volume, as witnessed by the fact that a large number of the academic disciplines usually regarded as engaged in the enterprise of cognitive science are represented herein. Indeed, even though some of the chapters (e.g. chapter 5) are somehow hybrid and could thus be seen as falling within more than one discipline, a natural way of sorting them out in that respect is as follows: neuroscience (chapters 4, 5, 6, and 7), linguistics (chapter 10), philosophy (chapters 1, 2, 8, 9, 11, and 13), and psychology (chapters 3 and 12). Of course, the relative predominance of philosophy on this list stems from the relative predominance of matters eminently foundational throughout the book, matters having to do with general questions concerning the relations that hold among the principal protagonists of the story about human cognition: mind, brain, language, world, and action.

It is worth mentioning at this point that, besides sharing a subject (broadly conceived), a common feature of the chapters in this volume is the fact that early versions of them were presented at the Lisbon International Conference on the Foundations of Cognitive Science at the End of the Century. Most

I am grateful to Chris Peacocke and Jim Higginbotham for their encouragement and help. I also wish to thank Peter Momtchiloff and Charlotte Jenkins, Philosophy Editor and Assistant Editor at Oxford University Press, for their cooperation, patience, and efficiency. This Introduction was written while I was on a grant kindly awarded by the Portuguese Foundation for Science and Technology (FMRH/BSAB/171/00).

people who had the chance to attend the Lisbon meeting in May 1998 would very likely refer to it as a memorable event, not only because of the friendliness of the environment and other traits of the same kind, but mostly on the basis of a feeling that something rather like a "meeting of minds" happened there on the occasion, among the several researchers contributing to the present collection. Indeed, the lively and cooperative discussion and exchange of views that marked most of the conference sessions, as well as the genuinely interdisciplinary dimension of the debates, very often generated real insight into some of the most interesting and difficult issues in the foundations of cognitive science. Although some of these aspects are extremely hard to capture in print, it is not unreasonable to think that what happened in the meeting has had some sort of benign bearing upon the final versions of the essays here included.

The Mind as a Processor of Information

How should one characterize the task of cognitive science in a way that would enable us to obtain an integrated picture of the disparate contributions in this volume, in which they are all part of a single theoretical enterprise? A common and convenient way is to define cognitive science as the scientific study of the mind and of its role in the production of intelligent (purposeful, goal-oriented) behaviour.

Although there is some controversy surrounding the issue, one should perhaps note that the term 'mind' is usually taken in this context to refer not only to the human mind, but in general to the mind of any intelligent information-processing system. Hence, even machines and artefacts of certain kinds—not to speak of animals—should not be ruled out, at least at the outset, as lacking mental activity in the relevant sense. As long as they can be reasonably counted as intelligent information-processing systems, they are assumed to be endowed with minds. (Whether there really are, or could really be, machines that were capable of meeting that condition is in itself a moot issue in the foundations of cognitive science, and one that need not bother us here.)

The leading idea underlying the above identification of the subject matter of cognitive science is that minds are basically processors of information; or, given that cognition is just (in one sense) information processing, it is the equivalent idea that minds are essentially cognitive devices. This idea seems to be rather pervasive in contemporary cognitive science. In fact, it even seems to constitute one of the few substantive foundational assumptions in the area that have been relatively immune to dispute so far. In particular, the assumption (or certain versions of it) is clearly endorsed both on classical or symbolic approaches to the mind, and on connectionist or non-symbolic approaches, these being the two main opposing methodological schools of thought currently available in cognitive science. Indeed, almost all those who work in this field would agree in identifying the processes to be studied in cognitive science as being those commonly involved in the information-processing activity of the mind (or of the brain), namely the processes of receiving, storing, retrieving, modifying, and transmitting information of various kinds. Accordingly, virtually everyone working in the area would be prepared to count as paradigmatic instances of information processing everyday phenomena such as visual perception and language comprehension, some important aspects of which are dealt with in chapters 12 and 7 of the present collection (respectively).

Thus conceived, the field of cognitive science covers not only cognition proper, the usual paradigms of which are mental events and processes such as conscious thoughts and inferences and intentional mental states like beliefs and desires, but also any other mental phenomena in which information processing happens to play a central role. Hence, in so far as they can be subsumed as instances of information-processing activity (which in many cases is clearly the case), a variety of mental states and events traditionally grouped under the heading of 'experience' and often contrasted *in limine* with cognitive states and events also fall within the scope of cognitive science. Among those states and events are notoriously sensations and perceptions, for instance visual experiences such as the experience of seeing a red object moving around in one's visual field; and auditory experiences such as the experience of hearing a piano sonata. In other words, both propositional attitudes—as philosophers call psychological states like belief and desire— and sensory experiences are part of the subject matter of cognitive science (note that on some views cognitive scientists should be interested in propositional attitudes just to be able to eliminate them from proper scientific inquiry). Yet, only states of the former kind are strictly cognitive in the following sense. In order to be in one of these states a system must possess and employ an appropriate set of concepts or ways of categorizing things; for instance, a system or organism cannot be in the state of believing that a cow is in front of it without having the concept *cow*, but it certainly can see a cow in front of it (a visual experience) without having such a concept.

So, one way to form a unified picture of the plurality exhibited by the chapters in this volume is to see them all as engaged in the task of explaining some aspect or other of the workings of the mind as an information-processing device. (As we shall see in a moment, this description is only partially accurate and needs qualification; but it will do for immediate purposes.)

Representation and Computation

The widely held assumption that the mind is basically an information-processing device is normally supplemented by two other relatively uncontroversial foundational claims. These claims complement that assumption by providing further specifications of what is going on as part of the information-processing activity. In particular, they can be seen as providing answers to the following two questions. (*a*) What in general is processed by the cognizing mind (or brain)? In other words, what is the nature of the information manipulated by the mind (or brain)? (*b*) What is the general form of the processing? What is it like and how does it proceed?

The first claim answers question (*a*) by specifying the objects over which the information-processing activity is defined as being essentially mental representations. Roughly, these are items in the mind or brain of a given system that in some sense "mirror", or are mapped onto, other items or sets of items; the latter are typically items external to the system: objects, events, situations, etc., in the world (typically, in the immediate environment of the system). Thus, it is assumed that cognition operates on mental representations, in the sense that these are basically what is being processed by the minds or brains of intelligent information-processing systems when they are performing given cognitive tasks. These tasks range from relatively simple ones, such as detecting the presence of a predator in the surroundings, to very complex ones, such as proving a mathematical theorem.

Note that mental representations might come in a wide variety of forms, there being no commitment in the claim itself to a specific kind of representation or to a particular sort of representational vehicle. According to taste, or theory, or purpose, mental representations might be thought of as images, schemas, symbols, models, icons, sentences, maps, and so on. Their job is to provide systems with the information they need to control their behaviour and guide their interactions with the environment. Mental representations are thus supposed to depict, by means of any vehicle that turns out to be appropriate to do that job, not only aspects and states of the outer world, but also aspects and states of the inner world, namely internal states of the system.

The second claim answers question (*b*) above by identifying the nature of the processing, by describing in general what goes on inside the "black box" (to use a familiar metaphor). The processing is characterized as being essentially computational, as consisting of a series of computations executed by the system (or by its mind, or brain). Given the first claim, it follows that the computations involved in cognition are defined over a set of mental representations. Roughly, this means that the cognitive tasks performed by the mind or brain characteristically consist in generating, in an effective way, certain mental representations as outputs, on the basis of certain mental representations given as inputs. The operations performed by the visual system or by the language faculty in humans are clear cases of computations in this sense: in the former case they take 2D patterns (retinal images) as inputs, and yield representations of 3D scenes as outputs; in the latter case they take acoustic representations of utterances as inputs, and yield semantic representations as outputs.

Now it might turn out that even these two foundational claims, as well as the underlying conception of the cognizing mind as a processor of information, come to be challenged at some point. Yet, as often emphasized in the literature in the area, they have so far enjoyed the special status of shared principles or background assumptions. In other words, they have been regarded as defining the whole enterprise of cognitive science. It comes then as no surprise that both claims are endorsed—even if tacitly in most cases—throughout the present collection, at least if one relies on appropriately generic formulations as the ones provided above. Of special interest in this context is the fact that the claims in question seem to cut across the methodological dispute between symbolic and non-symbolic models of cognition, a dispute that comes to the surface at several points along this volume.

An additional way of forming a unified picture of the variety of contributions and approaches contained in this book is to see them as somehow presupposing these two assumptions about the nature of the cognitive workings of the mind: the assumption that the mind represents and the assumption that the mind computes.

These foundational claims become controversial as soon as one tries to flesh them out, and be more specific both about the nature of the posited mental representations and about the nature of the computations defined over them.

Thus, one might want to hold the view that the mental representations involved in cognition are intrinsically symbolic or linguistic; that they are, to take a familiar proposal, sentences in a "language of thought". One might want to couple this with the sister idea that the computations involved in cognition are like the operations characteristically performed by digital computers, consisting in the application of purely formal rules or procedures to syntactically defined items (language-like mental representations). These views represent, however, specific methodological proposals, particular ways of fleshing out the above claims, and are thus far from being immune to criticism. Indeed, their joint endorsement partially defines the classical or orthodox approach to cognitive science, a view sometimes known as the epresentational Theory of Mind. Despite still being very influential, the classical approach is now highly controversial. As far as the present volume is concerned, it is explicit assumed in James Higginbotham's chapter (chapter 10), with respect to language understanding, and endorsed in Zenon Pylyshyn's contribution (chapter 12), with respect to visual representation and object tracking; but it is explicity rejected in the chapter by Ilya Farber, Will Peterman, and Patricia Churchland (chapter 4) with respect to spatial representation and knowledge.

Another well-known way of fleshing out those claims is given in the rival connectionist picture of the workings of the mind. On this view there is still room for mental representations, as well as for computations performed on them. But connectionist representations and computations are of a radically different ilk. For one thing, mental representations are here definitely non-symbolic and non-linguistic: they do not represent in the way that words and sentences represent. According to usual versions of connectionism, the mind's information-processing activity should be regarded as being intrinsically linked to, and impossible to divorce from, its physical medium: the brain. Thus, the computational processes involved in cognition are highly sensitive to the physical system in which they are implemented and cannot be assimilated to the purely formal operations performed by digital computers. Rather, connectionist computations proceed through the simultaneous operation of a large number of basic processing devices—the so-called units, nodes, or artificial neurons (the counterparts of the brain's neurons)—that interact with one another and form a complex multi-layered network. In contrast with standard classical or symbolic computations, which are both local and sequential, connectionist computations are distributed and parallel (at least on some branches of connectionism); semantic content and representational properties might accordingly be assigned simultaneously to several nodes, and so connectionist mental representations are also distributed and not local. Cognition is modelled on the basis of artificial neural networks, structures that mimic in a simplified way the complex operations and processes of the brain. In spite of being highly controversial too, this connectionist picture has been embraced or explored by many cognitive scientists, especially those working in neuroscience and cognitive neurobiology. As far as the present volume is concerned, the approach is explicitly endorsed in the chapter by Farber, Peterman, and Patricia Churchland (chapter 4) and in Paul Churchland's contribution (chapter 5), where cognitive skills, like those involved in spatial representation and moral discrimination (respectively), are modelled by means of neural networks and given a connectionist treatment.

Beyond Cognition?

Notwithstanding what has been said so far, recent developments seem to indicate that cognition is not in fact all there is to the mind that is of interest to cognitive science. This seems to be the case even if one adopts a liberal view about the scope of cognition, a view according to which cognition includes not only the realm of strictly cognitive or "conceptual" events and states, events and states endowed with propositional content, but also the realm of experience (sensation and perception).

The above remark has a certain air of paradox, since the modifier 'cognitive' employed in the term 'cognitive science' suggests a restriction to cognition, to the study of information-processing activity as it occurs in perception, memory, knowledge, language understanding, reasoning, problem solving, and so forth; that is to say, it suggests a restriction to the cognitive side of the mind. Nevertheless, the fact is that a gradually increasing focus has lately been placed within cognitive science upon what prima facie are non-cognitive mental processes, states, and events, especially those belonging in the so-called volitional and emotional departments of the mind. (We employ here for convenience the classical tripartite picture of mental life, a picture on which the mind is seen as divided into the separate areas of cognition, conation, and affect.) The paradox is only apparent, of course. For either the mental phenomena in question are in the end correctly describable as genuine instances of information-processing activity and turn out to be at bottom cognitive in nature; or an adequate redescription of the province of cognitive science is needed, a redescription which should *inter alia* include a revision of the conception of the mind as a processor of information.

At any rate, a considerable amount of research has been going on in cognitive science into the nature of a range of mental phenomena which cannot straightforwardly be reduced to, or do not centrally involve, cognition or information processing.

Emotion is a case in point. Emotional processes have in recent times been at the centre of a great deal of work in cognitive science, work carried out not just in those disciplines where there is a traditional concern with such phenomena—disciplines such as psychology or philosophy—but also in other disciplines, most notably cognitive neurobiology. One of the hot issues here is precisely the extent to which cognition is involved in emotion. Roughly, the debate in the area opposes cognitivist accounts, on which emotion is at bottom a species of cognition, and non-cognitivist accounts, on which emotion cannot be reduced to cognition. That cognition is present in emotion seems to be beyond doubt. As shown by Antonio Damasio in chapter 6, emotional processes invariably contain a clearly cognitive ingredient in the form of an evaluative judgement made by the subject. The problem is rather whether or not cognition, in the sense of information processing, is the essence (so as to speak) of emotion. One should note that the converse issue, the issue about the extent to which emotion is involved in cognition, is also a key issue in the area; it is extensively discussed by Antonio Damasio in his book *Descartes's Error* (A. R. Damasio 1994).

Consciousness, as represented by a range of higher-level mental states traditionally viewed as mysterious in their irreducibility to objective or scientific description and explanation, is also a case in point. Of special interest in this context is the form of consciousness Ned Block calls *phenomenal* consciousness (chapter 1), and in particular the set of sometimes associated properties known as *qualia*. These are the subjective qualities of sensory experience, vivid examples being the allegedly unique phenomenological aspects of the experience of having an orgasm, and the allegedly unique phenomenological aspects of the experience of listening to the sound of Galician *gaitas* (bagpipes). The extent to which consciousness in general, and phenomenal consciousness in particular, can be given a satisfactory account in terms of the information-processing model is also a matter for intense dispute.

This broadening of the scope of cognitive science, making room for mental phenomena that are not clearly cognitive in nature (even if they turn out to be so in the end), is well-represented in the present collection. Emotion and feeling are addressed by Antonio Damasio in chapter 6, consciousness by Ned Block in chapter 1, and morality in chapter 5 by Paul Churchland. Furthermore, if the main line of reasoning advanced by John Searle in chapter 13 is sound, then modelling human action in terms of information processing, in the sense of construing it as systematically arising out of some appropriate set of desires and beliefs, gives us a wrong picture of the relation between mind (volition) and action.

Foundational Issues

Apart from foundational work on methodological issues, including the discussion of background assumptions and guiding principles such as those considered earlier on, there are in the foundations of cognitive science—just as in the foundations of other scientific subjects—two main interrelated kinds of research. The first is concerned with the clarification and elucidation of the most basic or central concepts employed in the discipline. For example, the task of explaining such concepts as *information, knowledge, concept, consciousness, cognition,* and so on, belongs in this segment of foundational work in cognitive science. The second involves the investigation of a set of highly general and speculative issues about the nature of mind and cognition, issues that are motivated by developments occurring in the several disciplines contributing to cognitive science. These questions are predominantly philosophical in character and some of them, for example the so-called mind-body problem, have been discussed by philosophers for centuries. Thus, the age-old inquiry into the nature of consciousness, the question of whether it is possible to simulate mechanically intelligent behaviour, and the reflection on whether cognition and thought are possible without language, all belong in this segment of foundational work in cognitive science.

Foundational work of the latter kind ranges from research that is straightforwardly philosophical in nature, in which general questions are explicitly addressed, to research that, although not dealing explicitly with such problems, generates results that are highly relevant to their discussion. A related remark is that, by analogy with foundational work in other scientific disciplines, it is plausible to regard foundational work in cognitive science as being in most cases just a natural continuation of non-foundational work, the differences between the latter and the former being essentially differences in degrees of generality and scope, and not differences in kind. Accordingly, even though for obvious reasons philosophy should be seen as playing a pivotal role in the foundations of cognitive science, it is by no means accurate to view it as having the monopoly over research in the area. Indeed, it becomes apparent from the previous observations that important foundational work of any of the kinds mentioned above can be (and has been) carried out in any one of the other academic disciplines involved in cognitive science. A clear illustration can be found in this present collection. One can find conceptual work in the foundations of cognitive science (foundational work of the first kind) not only in Block's important distinction between two concepts of consciousness, *phenomenal*-consciousness and *access*-consciousness (chapter 1), but also in Antonio Damasio's important clarification of the notions of emotion and feeling (chapter 6); likewise, one can find "philosophical" work in the foundations of cognitive science (foundational work of the second kind) not only in Donald Davidson's defence of a broadly linguistic approach to thought and cognition (chapter 8), but also in Susan Carey's discussion of certain forms of cognition in infants and non-human primates and in her conclusion that they are radically non-symbolic and non-linguistic (chapter 3).

Three research themes have acquired a salient position in foundational work of the second kind (of course, there are natural intersections with foundational work of the first kind and with foundational work on methodological issues). They all relate to the connections between the mind, especially the human mind, and other intervening elements in any adequate account of its workings as a cognitive device. The first theme concerns the relations between the mind and the brain—or, to be precise, the relations between the mind and the central nervous system. A second theme concentrates on the relations between mind and language, by 'language' meaning either a natural language like English, or any other symbolic system of representation. Finally, a third group of issues concerns the relations between the mind and the world, including aspects about the evolutionary history of both and the connections between the mind and intentional behaviour and action upon the world.

These three research themes provide us with a convenient way of sorting out the principal contributions made by each author in the present collection to the foundations of cognitive science.

Mind and Brain

The first group of essays—chapters 1, 4, 5, 6, and 7—addresses mostly issues concerning the relations between mind and brain.

It is known that mental activity has some sort of physical implementation in the brain. Indeed, there is good evidence that mental states, events, and processes are directly correlated with certain states, events, and processes in the brain—or, in general, in the central nervous system—in the sense that they co-vary with the latter in regular and predictable ways. The search for neural correlates (as they have been called) for a wide range of mental phenomena has always been accorded a key position within cognitive science, especially within those disciplines—the neurosciences—that are particularly suited to the task. More recently, the focus has somehow been shifting from a concern with cognitive phenomena that are more or less familiar in that respect, like language processing or memory, to a concern with prima facie non-cognitive and much less familiar aspects of mental life, like consciousness and emotion.

The inquiry into the neural basis of mental states, events, and processes is regarded not only as being intrinsically valuable but also, on some non-symbolic views of mentality and cognition, as disclosing the nature or essence of these states, events, and processes themselves. These views assume, roughly, that the physical implementation of mental phenomena in certain systems in the brain is something that defines them (at least partially); they contrast thus with symbolic views, on which the neural realization of mental phenomena—the "hardware" on which they are actually "run"—is in general seen as largely irrelevant to their identity.

In chapter 1 Ned Block discusses the issue of the neural basis of consciousness. He begins by noting that, as employed by philosophers and cognitive scientists, the term 'consciousness' is ambiguous, as it is used to express different concepts of consciousness: phenomenal consciousness and access-consciousness. The former is identified with experience: one is phenomenally conscious if one is having an experience; sensations are typical cases of conscious states of this kind. The latter is defined thus: "a representation is access-conscious if it is actively poised for direct control of reasoning, reporting, and action" (p. 3); thoughts and propositional attitudes are typical examples of conscious states of this kind. The case is then made for the distinctness of the concepts by showing that it would be possible for access-consciousness to be instantiated without phenomenal consciousness being thereby instantiated and also, more controversially, that the latter might occur without the former. Yet, as one can see from pairs of concepts such as *water* and *H₂O*, distinct concepts may nevertheless be co-referential; so nothing would prevent phenomenal-consciousness and access-consciousness from being actually correlated with the same system in the brain.

Armed with this distinction Block goes on to examine two views, which he diagnoses as guilty of conflating the two notions of consciousness. One is John Searle's account of consciousness as not being necessarily involved in habitual, routine, and memorized activities, such as for example driving a car. The other is Francis Crick and Christof Kock's claim that the area in the brain associated with vision known as V1 is not part of the neural correlate of consciousness. Block contends that their argument faces the following dilemma. If 'consciousness' means access-consciousness, then the argument is trivial; if phenomenal consciousness is meant, then it is unsound. Either way, the argument does not succeed in establishing a the desired substantive conclusion. However, there are two positive claims made by Block. One is the claim that Crick and Koch's contention that V1 is not part of the neural correlate of consciousness may in the end be right, but on the basis of independent considerations that are nevertheless implicitly contained in their own work; the other is the claim, also extractable from considerations presented by Crick and Koch, that the two concepts of consciousness may not be co-referential as they may have different (though overlapping) neural correlates.

The nature of the processes involved in emotion and feeling and their neurobiological basis are discussed by Antonio Damasio in chapter 6. Damasio is in general sympathetic to William James's approach to emotion, especially to James's idea that the essence of the emotional process is a "sense of the body", an emotion being basically a perception of a set of physiological changes that take place in the body in response to a given situation or event. Nevertheless, James's account should be subjected to two important kinds of amendment. The first concerns the role of cognition in emotion. According to Antonio Damasio, James generally underestimates that role. In particular, in James's account no place is assigned to the clearly cognitive and non-automatic evaluative judgements that precede most cases of emotional response. Damasio argues that a mental appraisal of the significance of the stimuli that cause an emotion is in many cases a crucial component of the emotional process; it is notably present in most emotions we experience as adults. The second correction to James's account concerns the role of the body in emotion. Damasio's claim is that such a role is in a sense overestimated by James. In particular, there is no room in James's account for a neural mechanism that would be able to generate, without the intervention of the body (or else in a supplementary way), that state of awareness of body changes that is the distinctive mark of an emotion. Damasio argues that evidence clearly shows that such a mechanism is available, that the brain has in itself the resources to form an "as-if-body" state (p. 106), a state that depicts the body as if the body were being activated and modified in the way characteristic to emotion. Hence, contrary to James's idea, the interposition of the body in the emotional process is not necessary (even though it is a fact in many cases).

Within the highly complex phenomenon of emotion Damasio makes a subtle distinction between (*a*) emotion proper, which he identifies as the expressive element, the public manifestation of the phenomenon, and (*b*) feeling, which he identifies as the internal, subjective experience. The core of an emotion is given in a set of dispositions to respond that induce a collection of changes both in the state of the body and in the state of the brain, these emotional responses being preceded by the mental appraisal process mentioned above. The subject's representation of those body and brain states, which assumes the form of a complex set of mental images, is what constitutes the experiential component—the feeling. According to Damasio, feelings are thus a privileged means the brain has of providing us with knowledge of our body, not in the everyday sense of the term, but in the sense of "cognition of our internal milieu, visceral, and musculoskeletal states"; feelings are nicely described by him as "the first step in letting us mind the body" (p. 107). The presence of both non-cognitive (or automatic) and cognitive elements in the emotional process is paralleled at the level of the brain systems that constitute the neural basis of emotion. Thus, the traditional picture of emotion as being strictly connected with the limbic system and older brain structures, especially the amygdala and the anterior cingulate cortex, is replaced in Damasio's account with a richer picture that also makes room for brain systems that support the *cognitive* aspects of the emotional process; in this picture modern brain structures, in particular the neocortex, are seen as playing a key role.

Damasio's view is that emotion is not an excrescence, a redundant feature of mentality; it has a salient regulatory function and contributes in important ways to the overall success of the interactions between organisms and their environments. Another feature of the mind that is surely no less crucial in that respect, having a conspicuous "survival value", is spatial representation and reasoning. Indeed, a great deal of successful behaviour, behaviour that satisfies our needs and wants, is highly dependent upon our ability to represent space, to perceive the relative positions of independently existing things with respect to our body (or head) and to one another. This ability is involved in the execution of both very modest tasks, such as reaching for an object and grabbing it, and rather sophisticated ones, such as solving a geometry problem. As one can see from simple cases where the ability is exercised, forming an internal representation of space requires in general the joint use of sensory and motor information, as well as the integration of various sensory modalities (sight, touch, etc.). The nature and structure of spatial representation and reasoning, and how it relates to other forms of mental representation, have become a topic of great interest in the foundations of cognitive science.

In chapter 4, Ilya Farber, Will Peterman, and Patricia Churchland focus on the issue of how brains and nervous systems, especially mammalian brains and nervous systems, are able to represent space and carry out spatial reasoning. Their central thesis is that spatial representation, as it occurs in humans and other mammals, is fundamentally non-symbolic. This means that the way in which it happens to be physically implemented in the brain and nervous system, the specific configuration of the underlying neural systems, is a constitutive feature of our ability to represent space, something that determines the intrinsic nature of spatial representation. Given *inter alia* the important position this form of representation occupies in human cognition (for example, it is clearly involved in forms of self-representation such as the representation of one's own body), this thesis is intended to be part of a broader picture of mental representation as non-symbolic in the above sense. Accordingly, Farber, Peterman, and Churchland mount a vigorous attack on symbolic or linguistic approaches to cognition in general. Their main target is the familiar brand of symbolic approach known as functionalism. Functionalists assert that the nature of mental states can and should be described without any reference to the physical structures in which they are implemented; they are to be individuated rather in terms of their function within a given network of mental states.

In support of their non-symbolic approach to spatial representation Farber, Peterman, and Churchland draw on three kinds of empirical evidence: (*i*) behavioural data from animal psychology; (*ii*) data from neuroscience about the neural basis of spatial representation in mammals; and (*iii*) data from lesion studies, particularly studies of impairments of spatial representation and reasoning that result from damage to the parietal lobe. They argue that the symbolic view is immediately undermined when confronted with data coming from any of these sources. As to data of kind (*i*), Farber, Peterman, and Churchland take such data as indicating that spatial reasoning based on representations of object-centred space is carried out in organisms in which it seems absurd to postulate any relevant representational systems that are language-like or symbolic. Baboons, nutcrackers, bears, ravens, and rats are able to solve a host of problems—the hiding problem, the trapeze problem, the problem of finding or retrieving food, etc.—that require complex spatial reasoning, reasoning surely not different in kind from that which is carried out by humans. In particular, in order to solve such problems animals must be able to represent accurately the relation of their own bodies to various objects and other bodies in space. The proponent of the symbolic view is then confronted with the following dilemma: either she claims that there is indeed a difference in kind between human and animal representation of space, which is implausible; or she rules out spatial representation in general as being nonsymbolic, in which case a way would have to be found to discriminate between it and other cognitive skills of the same level of complexity. Data of kind (*ii*) are taken to show that the mammalian brain and nervous system is able to build representations of object-centred space in accordance with a radically non-symbolic model. The region of the brain that is responsible for producing "objective" representations of space is the posterior parietal cortex, and the crucial explanatory hypothesis invoked by Farber, Peterman, and Churchland is the Pouget-Sejnowsky Hypothesis. According to this hypothesis, the posterior parietal cortex generates *basis functions*, which compute such representations; a basis function is the product of two kinds of processing units (neurons), eye position units and retinal position units, such a product being computed by hidden units (interneurons in area 7 of the posterior parietal). According to Farber, Peterman, and Churchland, these results from neuroscience provide us with strong evidence against symbolic models of spatial representation. Finally, as to data of kind (iii), the authors claim that impairments in spatial reasoning, such as hemineglect, the tendency to ignore or neglect objects in particular regions of space, are hardly accountable in terms of symbolic approaches to spatial representation (whereas non-symbolic approaches have no difficulty in explaining them).

Language processing, the way the human mind is able to produce and understand spoken or written language, is a research area in cognitive science that has many significant implications for foundational inquiry into the nature of the cognizing mind and its relation to the brain. The neural basis of some central aspects of language processing is the issue addressed by Hanna Damasio in chapter 7. From the perspective of neuroscience, and working at the level of large-scale systems in the human brain, she deals with two sorts of cognitive tasks that are crucially involved in language production and comprehension: word retrieval and concept retrieval. The former consists in the ability competent speakers of English (say) normally exercise when, given a perceptually presented item (a particular person, an animal of a certain kind, an artefact of a certain kind), they recognize it by coming up with a correct word for it or for the kind it belongs to (a proper name like 'John Smith', a common noun like 'cat', or a common noun like 'hammer'). The latter consists in the ability we exercise when, given a perceptually presented item (e.g. a cat), we recognize it by describing it in an appropriate way, as having such and such salient features and properties (e.g. 'small feline that people keep at home...'). As Hanna Damasio points out, the fact that the former ability has not been exercised for some reason (one sometimes forgets the words) is not evidence that concept retrieval has thereby failed: one might have a concept for the presented item, as incorporated in some adequate description one might produce, without being able to produce the name. This is paralleled at the level of the corresponding neural systems: an impairment of the area of the brain that supports the retrieval of names for persons (the left temporal pole) may leave intact the concept-retrieval ability (see p. 110). Of course, word retrieval is taken as being in general sufficient for concept retrieval in the above sense.

Hanna Damasio investigates the neural systems underlying word retrieval and concept retrieval. She draws heavily on studies of two kinds, lesion studies and PET (Positron Emission Tomography) studies, conducted by her, in which recent sophisticated techniques for correlating cognition and brain are applied. These techniques are: functional imaging, by means of which brain activity is indexed when a subject is performing a certain cognitive task; and a modern version of the lesion method, by means of which hypotheses about the neural basis of cognitive processes are tested by considering damaged areas of the brain. It turns out that the results obtained in the lesion studies about the neural systems involved in word and concept retrieval largely coincide with the results obtained in the PET studies. The general conclusions Hanna Damasio draws from these studies are as follows. Firstly, the traditional anatomical map of language areas, or neural correlates of language processing, comes out as rather incomplete. According to such a map, language correlates in the brain are mostly restricted to two well-known areas (both located in the left hemisphere), Broca's area and Wernicke's area. Hanna Damasio argues that this picture is too simple and should be replaced by a more complex one, a picture in which many other regions of the brain (some of them in the right hemisphere) are involved in language production and comprehension; these regions are connected by bi-directional pathways and form a complex system. For example, the retrieval of names for particular persons seems to correlate with the left temporal pole, an area that is distant from both Broca's area and Wernicke's area. Furthermore, the neural correlates for the retrieval of concepts for persons, on the one hand, and for the retrieval of concepts for tools, on the other, seem to be situated in different hemispheres. Secondly, the neural correlates for word retrieval with respect to a given item are separable from the neural correlates for concept retrieval with respect to the item in question; word retrieval seems to correlate with regions in higher-order cortices of the left temporal lobe, whereas concept retrieval seems to correlate with regions in "higher-order cortices in right temporal polar and mesial occipital/ventral temporal regions, and in lateral occipital-temporal-parietal regions" (p. 116–17). Thirdly, if we focus our attention just on word retrieval, we see that neural correlates seem to vary according to the conceptual category to which the presented item belongs. The systems in the brain that support word retrieval with respect to persons, word retrieval with respect to animals, and word retrieval with respect to tools seem to be partially segregated from one another; the first systems appear to be located in the left temporal pole, the second systems in the left infero-temporal cortex, and the third systems in the posterolateral infero-temporal cortex. Similar results apply to concept retrieval, which also seems to vary according to the conceptual category of the presented object.

In chapter 5, Paul Churchland deals with the question of how the human brain and nervous system is able to produce such things as, for example, moral virtues (as well as vices, of course), and how the brain is able to generate higher-level mental functions such as those associated with moral representation and knowledge. Approaching the issue from a connectionist standpoint, Paul Churchland's central contention, for which he provides detailed argument and evidence, is that a vast number of significant phenomena involving morality can be given an integrated explanation in terms of neural-network theory. The basic idea is that the concepts and methods of connectionist cognitive modelling, which are employed to account for our performance of a host of natural cognitive tasks (such as visual perception or language processing), can be successfully applied to the moral realm too. Connectionist cognitive modelling proceeds by using simulated networks of simple processing units to explain human cognition. Such units resemble in a number of relevant respects individual neurons in the human brain, so that we thus obtain a model of the cognitive workings of the brain. Of course, the model is oversimplified: for one thing, neurons in living brains clearly outnumber processing units in artificial networks. But it gives us an overall correct picture of cognitive activity as consisting in complex interactions among large numbers of simple processing units.

Paul Churchland focuses on issues in the branch of moral theory known as metaethics. This is *inter alia* the study of moral cognition, of the nature of moral judgement and moral knowledge, and it includes questions about how these are acquired and exercised. A cardinal assumption he adopts, an assumption needed to warrant the extension of neural-network theory to the moral realm, is the view that moral cognition is essentially a set of skills, a set of complex perceptual, reflective, and behavioural skills that a morally knowledgeable adult possesses (p. 79). As such it differs only in detail from any other form of human cognition, most notably scientific knowledge. (One should note at this point that the parallel between moral cognition and progress, and scientific cognition and progress, plays an important role in the main argument of Paul Churchland's paper.) Hence, the formation and deployment of the set of skills embodying moral knowledge and representation should be, in principle, as capable of being modelled by means of neural networks as those skills embodying any other form of knowledge and representation.

Paul Churchland begins by highlighting the merits of a prospective research program based upon a systematic collaboration between the moral discipline of metaethics, on the one hand, and cognitive neurobiology, on the other hand. Such collaboration assumes the form of a bilateral interaction, with results from research undertaken at the micro-structural level contemplated by neural-network theory simultaneously feeding and being fed by high-level reflection in the domain of moral knowledge. The program is motivated by an analogy established with the interaction existing between cognitive neurobiology and other philosophical disciplines, especially the philosophy of science, an interaction Paul Churchland sees as theoretically fruitful and insightful. He then goes on to examine, from the point of view of neural-network theory, virtually all of the topics and issues commonly addressed in metaethics: moral knowledge, moral learning, moral perception, moral ambiguity, moral conflict, moral argument, moral virtues, moral character, moral pathology, moral correction, moral diversity, moral progress, moral realism, and moral unification. Moral knowledge, for example, is described (p. 81) in neural-network terms as being embodied in an intricate configuration of weighted synaptic connections; such connections partition an abstract space of possible activation patterns of neurons or processing units into a hierarchical set of prototypical moral categories ("morally significant action"/ "morally non-significant action", "morally good action"/"morally bad action", etc.).

Paul Churchland sees the general account that emerges from such an application of connectionist cognitive modelling to the above moral phenomena as belonging in a particular pre-existent tradition or school of thought in metaethics, a tradition known as virtue ethics. Virtue ethics is a family of views about the nature of morality and moral judgement that go back to Aristotle and have been recently endorsed by ethicists such as Mark Johnson,

Owen Flanagan, and Alasdair MacIntyre. The leading idea of virtue ethics is that morality is to be fundamentally explained, not in terms of some fixed set of ultimate principles or rules that would govern moral judgement, but in terms of inner characteristics or virtues that individuals gradually acquire on the basis of complex dealings with the social environment. According to Paul Churchland, the account of morality provided by cognitive neurobiology fits nicely with virtue ethics. However, in spite of recognizing a large area of agreement, he closes his chapter with a critical examination of the treatment given to the issue of moral progress by virtue ethics theorists Flanagan and MacIntyre. Both these writers endorse, for different reasons, a frankly sceptical view with respect to moral progress, claiming that there are no real advances in the sphere of human moral consciousness. Paul Churchland argues that such a scepticism is misplaced by drawing on the already noted parallel between moral cognition and scientific cognition, a parallel supported by the findings of neural-network theory; since moral knowledge does not differ in kind from scientific knowledge, to the extent that there is genuine progress in the domain of science, to that very extent there is also genuine progress in the moral realm.

Mind and Language

A second group of contributions—chapters 3, 8, and 10—can be seen as dealing mainly with issues concerning the relations between mind, especially cognition and thought, and language.

Questions about the mind/language connection, for example questions about the extent to which human thought and cognition are constitutively determined by their usual verbal expressions in a natural language, are nowadays best addressed from a multi-disciplinary perspective. They have always been of interest to philosophers, occupying a prominent position in current debates in philosophy of language and in philosophy of mind. Recently they have also attracted the attention of many researchers working in other branches of cognitive science, particularly linguistics, psychology (cognitive and developmental), anthropology, and computer science; a number of significant results obtained in these disciplines have helped shed new light onto a number of traditional problems about the nature of the relation between human cognition and thought, on the one hand, and their linguistic clothing, on the other.

The mind/language connection is very often thought of as bi-directional, as consisting in some sort of mutual dependence between its terms. Yet, this way of looking at the connection may naturally give rise to conflicting views. Taken in the mind-language direction, the connection seems straightforward. Language processing, the ability to learn, use, and understand a natural language, is in some sense a mental activity. The key concept here is meaning, a concept under investigation in many areas of cognitive science. Indeed, the meaningfulness of otherwise meaningless sequences of sounds or marks is something that the mind somehow imposes upon language. But the specific role assigned to the mind in that respect, as well as the nature of the mental and cognitive machinery involved in language production and comprehension, are matters of intense dispute. On some accounts cognition and thought are claimed to be prior to language either metaphysically or in the order of theoretical explanation. These are different, independent priority claims. The priority of thought in the latter sense usually means that linguistic meaning should be ultimately explained in terms of mental states, complex beliefs and intentions in the minds of speakers and hearers; in the former sense, it usually means that there could be thought without language. Taken in the language-mind direction, the connection is sometimes captured in the claim that cognition and thought are shaped by language in the sense that the availability of thoughts of certain kinds is dependent upon the availability of certain linguistic resources adequate to express them; in other words, the claim is that there are thoughts we are able to apprehend or express but would not be able to apprehend or express if we lacked a language endowed with such resources. Likewise, on some accounts language is claimed to be prior to cognition and thought either metaphysically or in the order of explanation. The priority of language in the latter sense usually means that our capacity for entertaining thoughts, especially our capacity for having propositional attitudes such as beliefs and wishes, should be explained ultimately in terms of certain relations holding between us and linguistic items such as sentences of a natural language; in the former sense, it usually means that thought is not possible without language.

The question of whether language is prior to thought in the metaphysical sense receives an explicit positive answer in Donald Davidson's contribution (chapter 8) and an implicit negative answer in Susan Carey's contribution (chapter 3). Of course, a genuine theoretical conflict would arise here only if roughly the same thing were meant by the term 'thought' on both views, which is far from being clearly the case.

Carey takes as her starting point a certain set of cognitive primitives, concepts that are basic ingredients in our conceptual scheme or system of representation of the world. She considers two key kinds of basic representational resources: concepts of object and concepts of number. The former include the concept referred to in the psychological literature as the *object* concept, in the sense of our general concept of a *bounded physical object* or *concrete particular*; and also representations of numerical identity and distinctness between objects—one and the same object versus two distinct but physically similar objects—as given in basic quantifiers such as *one, another*. The latter include our representations of the series of natural numbers, especially the first integers 1, 2, 3, and the associated counting system. These concepts are expressed in several ways in the lexicon or syntax of any natural language and play an important role in our overall representation of the world.

The crucial foundational questions Carey raises in her paper about the above set of cognitive primitives are as follows. Are these concepts language-dependent, in the sense of presupposing a prior representational system with the structure and resources of a natural language? Are the associated cognitive abilities constitutively linked to the language capacity? Are they necessary parts of the language acquisition device? These questions are addressed from the broader perspective of a general research program in comparative cognition, a program that has been carried out by Carey and Marc Hauser and whose results are reported in this chapter. In such studies the presence or absence in infant human cognition of given representational resources and abilities is compared with their presence or absence in non-human primate cognition. The underlying idea is that by identifying the cognitive features in question both from an ontogenetic and from an evolutionary standpoint one obtains a clearer grasp of their nature and of their connection with the language capacity. Thus, two questions have to be answered with a view to arriving at a satisfactory answer to the above questions. First, do prelinguistic human infants have concepts of object and number of the sorts under consideration? Second, are these representational resources available to non-human primates? Carey argues that evidence arising from research in comparative cognition indicates that both questions should be given affirmative answers, and hence that our concepts of object and number should be viewed as parts of a core of innate knowledge; such knowledge is claimed to be ontogenetically and evolutionarily prior to the system of symbolic representation embodied in our use and understanding of a natural language.

Carey takes the *object* concept as a sortal concept, a concept that provides us with criteria for individuating and counting the entities falling under it. The individuation principles associated with our concept of a physical object include: (*a*) principles about object permanence, particularly the idea that material objects are expected to continue to exist independently of us as time goes by and when perceptual contact with them is lost; (*b*) spatio-temporal principles, especially the idea that two material objects cannot occupy the same portion of space at the same time and the idea that there must be a continuous spatio-temporal path linking successive appearances of one and the same material object. These and other principles are said to be constitutive of our *object* concept. Drawing on empirical studies where the technique known as the *looking time* method is applied both to preverbal human infants and to non-human primates, Carey contends that creatures of both kinds can be credited with the concept of a physical object as defined by these features.

Contrary to Jean Piaget's account, on which the *object* concept is intrinsically linked to the language capacity, Carey takes research carried out by Elisabeth Spelke, Karen Wynn, and others to show that the *object* concept, as well as the basic quantificational concepts *one* and *another*, are already available to prelinguistic human infants — babies with ages ranging from 2 and a half months to 3 months. Infants are described as knowing that material objects continue to exist when behind barriers and as making use of spatio-temporal principles such as the ones above for individuating material objects. A sharp contrast is drawn between the case of the sortal *object* and other object concepts, on the one hand, and the case of specific sortals, concepts such as *book, bottle,* and *doll,* on the other; indeed, Carey argues that human infants can only employ the latter range of concepts when the language device is already in place. Furthermore, in a series of studies undertaken by Carey and Hauser the looking time technique was also applied to non-human primates, first to wild rhesus monkeys and then to cotton-top tamarins, creatures that are even more distant from us in evolutionary terms. The results obtained turn out to be analogous to those relative to infant cognition: non-human primates of both species are described as possessing the sortal *object* concept and the concepts of numerical identity and distinctness associated with the quantifiers *one* and *another.*

As to concepts of number, Carey discusses three proposals concerning the nature of the system used by human infants to represent the first three integers 1, 2, 3: the Numeron List Proposal, the Accumulator Proposal, and the Object File Proposal. She interprets the literature in the area as providing us with evidence in support of the Object File Proposal, an account that distinguishes itself from the other two accounts in virtue of being definitely non-symbolic; unlike them, the Object File Proposal is not based on the idea that a distinct symbol should be assigned to represent each integer. Carey concludes that "there is no evidence for a prelinguistic representational system of the same structure of natural language count sequences, such as '1, 2, 3 . . . ' " (p. 43). However, she is inclined to favour the Accumulator model with respect to the system employed by non-human primates to represent integers.

The question of whether speechless creatures can or could have concepts and think is a highly controversial issue in the foundations of cognitive science. As we have seen, from Carey's viewpoint there is a range of basic concepts that are available to preverbal human infants and to nonhuman primates; it seems that such speechless creatures can be credited on that basis with thought and cognition.

A different and *prima facie* opposed approach can be found in Davidson's essay. Davidson investigates the requirements of thought: the conditions that an object—a system, a device, a creature—must in general satisfy in order to be correctly identified or recognized as a cognizing thing, in the sense of something that has concepts and is able to employ them in thought. He famously argues, here and elsewhere, that only systems or creatures that are very similar to us (mature human beings) can be legitimately credited with concepts and thoughts. One of the aspects that is taken by Davidson as crucial in establishing such similarity is precisely speech, mastery of a natural language; hence, on his view, speechless creatures in general and animals in particular do not literally have any concepts and are not literally capable of entertaining any thoughts. Of course, we may feel tempted to describe them as cognizing things or thinkers; but such ascriptions are only projections that we make on the basis of our own case, they should not be taken literally.

The gist of Davidson's main argument towards that conclusion is as follows. In order for a system or creature to possess concepts, and exercise them in thinking, it must have propositional attitudes: beliefs, desires, intentions, and so on. Propositional attitudes are something we ascribe by essentially using 'that'-clauses to specify propositional contents, for example when we classify a creature as believing *that there is food in the fridge* or as wishing *that the rain would stop.* Now, notwithstanding appearances to the contrary, animals and other speechless creatures are arguably incapable of having any propositional attitudes: no such attitude report can be literally true of them. Therefore, they lack concepts and thought.

In fact, Davidson invites us to equate having concepts and thinking with having propositional attitudes. Concepts are ways of sorting out items in the world, and to have a concept is to be able to, or to be disposed to, class (or not) a given item under the concept. But to be capable of recognizing an item as instantiating (or not) a concept is to be capable of judging or believing that it falls (or does not fall) under the concept; in other words, it is to be capable of having propositional attitudes. A difference should be discerned between recognizing an *F* simpliciter—in the sense of being able to, or being disposed to, discriminate *F*s from non-*F*s—and recognizing an *F* as such (as an *F*); the latter ability, but not the former, requires a creature to be able to, or to be disposed to, judge or believe that the item in question is an *F.* Davidson dismisses in this way the temptation we often have to ascribe concepts and thoughts to animals on the basis of their sometimes rather strong powers of discrimination, on the basis of their apparent ability to distinguish between items that do and items that do not fall under a concept; as he puts it, this ability consists of nothing other than mindless dispositions to respond in specific ways to items that *we describe* as instantiating the concept. Although it can surely be said of animals and other speechless creatures that they are able for example to *see* food in the fridge, it cannot be said of them that they are able to *see that* there is food in the fridge.

Davidson introduces three different but related sorts of requirement for having concepts or propositional attitudes; it turns out that each of these conditions can only be met by creatures endowed with a representational system with the structure and properties of a natural language. The first requirement is based on Davidson's holism about meaning and concept ascription. Concepts are individuated, not only in terms of certain relations they bear to items in the world, but also in terms of certain relations they bear to other concepts; some of the latter relations are entailments, others are relations of evidential support. Thus, one cannot possess a concept or have a belief without possessing a host of other concepts in a network of interrelated concepts or having a host of other beliefs in a web of beliefs. In order to be correctly classified as a thinker a creature must therefore be in possession of a relatively sophisticated conceptual repertoire, a system very much like the one embodied in a natural language. The second requirement is that in order to have concepts a creature must not only be in a position to make occasional mistakes in applying a given concept, for instance judging that a rat is crossing the street while watching a squirrel doing it, but also be in a position to recognize that the creature itself has made a mistake, which is clearly a higher-order propositional attitude. As Davidson puts it, "a creature that cannot entertain the thought that it may be wrong has no concepts, no thoughts" (p. 126); again, this involves crediting thinkers with sophisticated concepts such as the concepts of *error* and *objective truth*. Finally, a third set of requirements is imposed by concept possession upon the structure of thought and upon the structure of any language adequate to express it. Let us mention the most basic of such conditions: (*a*) the creative property of concepts and thoughts, i.e. the fact that concepts can apply to an endless number of items and the fact that there are infinitely many thoughts to be entertained, requires that such a language contain demonstrative terms and truth-functional connectives; (*b*) assuming that one cannot have thoughts without a framework for objectual reference and without possessing the general concept of an object, the language should also contain resources adequate to play the role of variables and quantifiers in the usual symbolism of quantification theory (note the contrast between such a concept of an object and the *object* concept envisaged by Carey in chapter 3).

In chapter 10 James Higginbotham focuses on semantics, the study of linguistic meaning, and its relation to the enterprise of cognitive science. Meaning is a key concept in the study of the mind/language connection. Indeed, an important feature of language processing, our ability to use and understand a natural language, consists in assigning meanings to spoken or written words and sentences of the language on particular occasions; and this is in some sense a mental activity, something that our minds do. How should word and sentence meaning in general be explained? How should one characterize understanding, knowledge of meaning? Aspects of the twin topics of meaning and understanding are the subject matter of Higginbotham's chapter.

Higginbotham approaches the topic of meaning from the standpoint of referential or truth-conditional semantics. According to this view, meaning is to be centrally explained in terms of reference and other related extensional notions, such as satisfaction and truth; all these are language-world relations, holding between linguistic expressions and items or sets of items in the world. Aspects of language use that apparently cannot be accounted for solely in terms of referential properties, for instance racial epithets and euphemism, are nevertheless only understood against the background of reference.

According to referential semantics, the meaning of a declarative sentence is given in terms of its truth-conditions, as these are compositionally determined on the basis of referential properties of lexical elements occurring in it and its syntax—the specific mode of combination of such elements in the sentence. The nuclear part of an account of meaning for a natural language takes the familiar Tarskian shape of an axiomatized body of statements, a *theory of meaning* for the language. The axioms of the theory specify referential properties for the primitives of the language, as well as the semantic significance of the available modes of combination; the theorems of the theory state truth-conditions for arbitrary sentences in the language according to their structure and the referential properties of their constitutents as specified by the axioms.

The issue Higginbotham then addresses concerns the place that should be assigned to referential semantics within cognitive science. He assumes a familiar view of cognitive science, on which the mental states studied therein are computational on the following sense. They are individuated in part by their content or representational properties, and in part by their causal powers (i.e. their interactions with other mental states and behaviour), the computations involved in the latter being essentially symbolic, i.e. defined over a purely formal basis. The connection between referential semantics and cognitive science so construed is captured in the following two claims. First, understanding or knowledge of meaning consists in genuine propositional knowledge, although mostly tacit, possessed by articulate speakers of a natural language; it is *knowledge that* as contrasted with *knowledge how*. In other words, there is such a thing as semantic *competence* in Noam Chomsky's sense of the term: a system of internalized representations of semantic rules and principles. Secondly, the objects of such knowledge are conditions on reference as given by the statements of referential semantics; for instance, able speakers are said to have tacit knowledge of axioms stating the denotation of singular terms of the language. Knowledge of meaning is thus modelled as a complex computational mental state, the inputs to which are the statements of referential semantics as embodied in a meaning theory as outlined above and the outputs of which are "behaviour and adjustments of states that go to exemplify our rationality in the use of language" (p. 147).

These two claims play the role of major premises in the account of meaning and understanding proposed by Higginbotham. He argues indirectly in their support by examining a set of three alternative proposals that have been influential in semantic theory and by identifying their shortcomings when compared to referential semantics.

The first proposal derives from deflationary or minimalist accounts of reference, as well as of the other semantic notions (satisfaction, truth) that are central to referential semantics. A consequence of such views is that the notion of reference is inadequate to play any explanatory role in an account of linguistic meaning. At least in the version considered by Higginbotham, which he attributes to Paul Horwich, word and sentence meaning are explained rather in terms of concepts and propositions expressed, which are essentially extracted from the way we use words and sentences. According to Higginbotham, the proposal founders because it is incapable of providing us with a satisfactory account of the significance of even such a simple sentential mode of combination as the predication schema; he argues that the proposal is unable to explain how we are able to know the meaning of a monadic predication on the basis of knowledge of the meanings of its components and appreciation of its syntactic structure. The second proposal rejected by Higginbotham is given in representationalist accounts of meaning and knowledge of meaning, on which the objects of tacit knowledge are taken to be mental representations; he claims that such views turn out to be inconsistent with the nature of linguistic competence and even with the practice of linguistics itself. The third proposal is encapsulated in Jerry Fodor's semantic views, especially in his thesis that, strictly speaking, natural languages have no semantics; only mental representations, words, and sentences in the language of thought have a semantics (in fact, a referential semantics). Higginbotham discerns three independent ingredients in Fodor's views: (_a_) the thesis that to understand a sentence of a natural language consists in mapping it onto a thought, the thought it conveys; (_b_) the thesis that such a mapping consists in translating the sentence into a sentence of the language of thought, a sentence which expresses the thought in question; and (_c_) the thesis that learning a natural language is learning how to map its sentences onto thoughts. Higginbotham argues that claims (_a_) and (_b_) are in the end consistent with referential semantics construed as a theory of knowledge of meaning; it is rather thesis (_c_), and the underlying idea that knowing a language is a practical ability, that should be rejected.

Mind and World

The dominant theme for a third and last group of chapters is the connection between mind and world, a connection of lasting theoretical interest to researchers working in cognitive science and its foundations. The mind/world connection is multifarious and can be unravelled into a number of different relations. It is convenient to use the familiar notion of a direction of fit to introduce some of these relations. One of them is knowledge, under which are subsumed several other central mind/world relations such as perception and memory; knowledge, as well as each of its species, can be described as having the mind-to-world direction of fit, in the sense that it is the mind that has to adapt to the world in order to produce knowledge. Aspects of knowledge are the subject matter of Zenon Pylyshyn's chapter (chapter 12), in which issues involving visual perception are investigated, and of chapter 11, in which Christopher Peacocke addresses issues about the integration of epistemology and metaphysics within a general theory of concepts. Another central relation between mind and world is action. Action can be described as having the world-to-mind direction of fit, in the sense that it is the world that adapts to the mind when it is acted upon by us and other creatures; issues involving the explanation of action and intelligent behaviour are discussed in John Searle's chapter (Chapter 13).

At a more general level than knowledge, a mind/world relation that has always been at the heart of philosophical reflection, especially in philosophy of mind, is the relation known as Intentionality or Aboutness. (Action is also intentional, but in a different and more usual sense). Many mental states and events are said to be intentional in the sense of being about, or being directed upon, non-mental things; these are typically things in the world: specific objects, events, or situations. Take the mental state someone might be in when she believes that London is pretty; such a state is said to be intentional both in the sense of having an intentional object, the city of London itself, and in the sense of having an intentional content, that London is pretty. The same goes for a host of other mental states and events such as desires, fears, thoughts, regrets, and so on. Intentionality cannot obviously be defined in terms of direction of fit: knowledge is an example of an intentional state with the mind-to-world direction of fit, and desire an example of an intentional state with the world-to-mind direction of fit.

Aspects of the problem of intentionality or meaning (in one sense of 'meaning')—the problem of explaining how there can be things in the mind/brain, presumably worldly things, that are about other worldly things—are explicitly addressed in Margaret Boden's chapter (chapter 2) and in Daniel Dennett's chapter (chapter 9).

Dennett launches a sustained criticism of the general picture of intentionality defined by two methodological claims, which he regards as having been omnipresent within the foundations of cognitive science. He labels these claims the "*content capture*" *assumption* and the "*isolated vehicles*" *assumption*. The first is a claim about content and (roughly) states that intentional content should be captured in terms of propositions or intensions; although Dennett has in mind mostly linguistic or symbolic representational resources, the claim is taken by him in a broader sense to mean that intentional contents should be directly specified in terms of data structures, a term he employs to cover not only linguistic items like sentences in the language of thought, but also representational devices like images, icons, and maps. The second claim is about the vehicles of intentional content and (roughly) states that it is theoretically possible, as well as desirable, to isolate such vehicles from the "outside" world; in particular, with respect to creatures like us (i.e. embodied nervous systems), there is an important level at which the bearers of intentionality—the *things-about-things* (as Dennett calls them)—can and should be taken in complete abstraction from their connections with the body.

According to Dennett, these two assumptions about intentionality have been rather pervasive in cognitive science. They owe their pervasiveness in part to their original status as useful idealizations or oversimplifications; like any other science at its initial stages of development, cognitive science needs some such oversimplified assumptions in order to get off the ground. However, these assumptions have paved the way for a set of foundational questions about intentionality that are, on Dennett's view, the wrong questions to ask at the current stage of inquiry. One such question is the ontological issue about what sorts of things, what kinds of items in the mind or in the brain, are the bearers of intentionality, the things-about-things. Are they mental represent-ations, for instance sentences in the language of thought, or else non-symbolic items like icons or maps, or are they just complex abilities generated by structures (neural networks) in the brain? From the standpoint of a broadly evolutionary account of cognition and intentionality, an account on which intentional properties of mind or brain components are basically explained in terms of their evolutionary functions, Dennett argues towards the conclusion that these two enabling assumptions should be put aside as they give us a distorted picture of the intentionality present in embodied nervous systems.

Dennett claims that the "content capture" assumption invites us to ask the wrong kind of questions and leads us in the wrong direction. In a nutshell, his argument is that describing the intentional contents of given pieces of cognit-ive machinery in the mind/brain by specifying certain propositions or certain concepts, the propositions believed or the concepts entertained, does not provide us with any idea of how these mind/brain components are able to carry out their intentional role; it does not give us any idea of how they are able to indicate, or contain information about, particular objects, properties, or events in the environment. In other words, content descriptions of that kind, the model of which is given in explicit expressions of intentional contents, do not provide us with an adequate basis for an explanation of intentional properties; they are bound to be incomplete in this respect. Intentional content should rather be approached in terms of functional descriptions, descriptions of the functions realised by the things-about-things present in embodied nervous systems as such functions are determined by the evolutionary history of the systems. Content specifications given on the basis of the propositional model are utterly unable to do the work; at best, they give us only mnemonic labels for functional structures, labels that merely allude to intentional roles performed by things-about-things.

Appealing as it may be, the "isolated vehicles" assumption also conveys a misguided picture of the mind and its intentional properties. Dennett describes the cardinal idea behind the assumption by means of a suggestive image, an image in which the mind is depicted as a control system and the body as the system controlled by it. Just as the control systems of many familiar machines can be isolated from the controlled systems, so that the latter systems can continue to operate even upon complete replacements of the former systems (for instance remote controls of TV sets can be replaced without any loss of function); so the mind, although as material as the body, can be isolated from the body in the sense that its central activity as a processor of information is insensitive to the physical medium in which the information is transmitted, processed, and stored (in particular, such activity is not affected by the physics and chemistry of the body). Of course, there must be points of contact between control system and controlled system, channels of information that link the mind to the body and to the "outside" world. They are of two restricted kinds, transducer or input nodes, and effector or output nodes, and their physical composition is surely relevant. Yet, according to the "isolated vehicles" assumption, it is possible to segregate transducers and effectors by regarding them as parts of the controlled system, the body, not of the control system, the mind.

Dennett thinks this is in many respects a seductive picture. Nevertheless, the envisaged segregation, as well as the underlying conception of the mind as an isolated and "clean" control system, fails when applied to creatures like us, and other embodied nervous systems. First, unlike the control systems available in many human artefacts, it turns out that minds, in other words the control systems for animals, are shaped in crucial ways by evolution and natural selection; in particular, minds are bound to interact and collaborate with control systems that happen to be much older than them in terms of evolutionary history, such as for instance hormonal systems. According to Dennett, facts of this ilk strongly militate against the envisaged segregation of bodily systems. Secondly, as Dennett aptly puts it, "evolution embodies information in every part of every organism" (p. 139). There are aspects of the interaction between mind and body that suggest that the picture of the mind as the control system of a body could even in a sense be reversed. Indeed, there is evidence that very old bodily systems sometimes guide the central nervous system and play a crucial role in controlling cognitive processes. An interesting case Dennett introduces and discusses is that of the role played by the emotional states we might loosely call boredom and interest in controlling the performance of cognitive tasks by young children; these states are crucial in guarding the cognitive systems of the children against debilitating mismatches. Dennett's supposition is that there is a very old bodily system, associated with visceral reactions to frustration, which underlies a capacity for boredom or interest and accomplishes the control in question.

Intentionality (or meaning) and evolution are also discussed in Margaret Boden's chapter, but in the different context of an examination of the connections holding between cognition (or intelligence) and life. This is an important and hot topic in the foundations of cognitive science, naturally related to issues and developments in the disciplines of AI (artificial intelligence), on the one hand, and A-Life (artificial life), on the other. Boden addresses two interrelated questions in her chapter. (1) What is the relation between cognition—as exemplified in perception, knowledge, language, and reasoning—and life? More specifically, is life presupposed by cognition, is it a distinctive mark of cognition, so that only living things could be cognizing things? (Note that these are intended as modal questions in the sense that even if, as a matter of fact, there are no non-living things that are capable of cognition, this would not yield affirmative answers to them. The same holds with respect to the questions below.) (2) Could human artefacts, for instance computers or other sophisticated systems, have life or cognition? Actually there are two different questions here: (2*a*) Could there be artefacts that were living things? (2*b*) Could there be artefacts that were capable of cognition?

On the one hand, Boden argues that whereas question (2*a*) should be answered negatively, thus contradicting the thesis defended by a number of people working in A-Life, question (2*b*) should be answered positively, thus vindicating the thesis defended by a number of people working in AI. A consequence of the conjunction of these two claims is that question (1) should receive a negative answer as well. On the other hand, Boden discusses in detail an argument adduced with a view to giving a positive answer to question (1), an argument that depends on the view that meaning and intentionality presuppose evolution.

Boden begins by adopting a characterisation of the notion of life whose essentials go back to Aristotle. According to this notion, the following features define life in the sense that a thing or system must possess each of them in order to be a living thing or system: self-organisation, emergence, autonomy, growth, development, reproduction, adaptation, responsiveness, and meta-bolism. She then contends that, among these constitutive features of life, self-organisation should be seen as central as it entails any of the other features on the list; indeed, each of them can apparently be extracted from the definition of self-organisation and can thus be seen as a variety of self-organisation.

The connection between life and evolution, particularly the issue of whether evolution should be added to the above list of features, is considered in the context of the already mentioned argument towards the conclusion that life is essential to cognition. The premises of the argument are the following claims (crudely stated): (*a*) meaning and intentionality are necessarily involved in cognition; (*b*) evolution is necessarily involved in meaning and intentionality; (*c*) evolution is a criterion of life (note that reading this claim as a sufficient condition and taking evolution as sufficient for life would be enough to run the argument). Boden accepts claim (*b*) on the basis of a discussion of several accounts of meaning and intentionality. Claiming that naturalistic accounts of meaning should be in general preferred to non-naturalistic ones, she goes on to consider two such accounts: informational semantics, on which meaning and intentional properties possessed by a system or thing are explained in terms of reliable causal correlations between worldly items and internal representations in the system; and evolu-tionary semantics, on which they are explained in terms of traits of the evolutionary history of the species of the system or thing. She argues that evolutionary semantics, especially the version that has been proposed by Ruth Millikan, is the best naturalistic account of meaning available. It follows from such an account that, as stated in claim (b), only systems or things that have evolved are capable of having meaning or intentional properties. Hence, taking premiss (*a*) for granted and endorsing (*c*), evolutionary semantics is committed to the thesis that only living systems or things are capable of cognition.

As a result of her discussion of questions (2*a*) and (2*b*), Boden comes to reject this thesis. As to question (2*b*), she dismisses an argument towards an affirmative answer, an argument against the possibility of ascribing cognition to human artefacts. The argument is inspired by Searle's attack against strong AI: artefactual systems, for instance AI-programs, are incapable of cognition because they necessarily lack genuine intentionality or meaning and meaning is presupposed by cognition (this is claim (*a*) above). According to Searle, such systems lack meaning because they lack the appropriate kind of inter-action with the environment. Boden's reply is that it is not clear that this argument applies to classical robots, and it is even less clear that it applies to evolved, embodied artefacts; in both cases the systems are capable of entering into real causal interaction with the environment. On the other hand, she also rejects a different argument to the same effect, the argument that artefacts cannot have cognition because they are unable to evolve and evolution is essential to cognition. Boden rejects this argument by rejecting its first prem-iss. Indeed, she argues that human artefacts can be genuinely evolved and mentions a few actual examples: certain creatures that populate virtual worlds, like the inhabitants of the software 'Tierra', and certain material robots, some of which are not only evolved but also embodied. As to question (2*a*), she argues towards the conclusion that human artefacts cannot be alive. Of course, this conclusion is not drawn on the basis of the claim that they cannot evolve; for, as we have seen, she does not accept this claim. Rather, Boden's argument is that human artefacts, even in the case of both embodied and evolved robots, necessarily lack one of the constitutive features of life, namely metabolism.

In his contribution (chapter 13) John Searle deals with issues involving the nature of practical reason, the relationship between rationality and delibera-tion, and the explanation of human intelligent behaviour. On the one hand, Searle's chapter contains an impressive attack against what he calls the Classical Model of rationality in action, a general conception of rationality and its connection with decision-making that has been largely influential in the philosophical tradition. On the other, the constructive part of his argu-ment consists of a general account of desire-independent reasons for action and their role in practical reason.

According to the Classical Model, human action is invariably explained by means of the belief-desire model, that is to say, in terms of an appropriate set of interrelated beliefs and desires which the agent has on the occasion; such beliefs and desires not only cause the agent's course of action but also explain it—are reasons for it. Thus, there is a straightforward connection between the explanations favoured by the Classical Model and the explanations that are familiar from so-called Folk Psychology, explanations whose theoretical status and value have been widely discussed in the foundations of cognitive science. Here is an example of an explanation of the sort. I suddenly decide to stand up and go to the kitchen *because* I believe there are bottles of mineral water there, I want to drink mineral water, and I believe that I will not get mineral water unless I go to the kitchen; these beliefs and desires, as well as a number of other beliefs and desires that would need to be invoked in a complete description, are taken as reasons for my purposeful behaviour.

Desire occupies a central position in the account of rationality in action embodied in the Classical Model. On the one hand, the presence of desires—construed broadly so as to include wishes, inclinations, preferences, etc.—is needed to generate action (belief alone is insufficient); on the other, rational deliberation consists in selecting the course of action that one recognizes as best satisfying our desires. From Searle's perspective, essential to the Classical Model is not (or not only) the claim that desire-dependent reasons can be reasons for action, which merely states a sufficient condition; it is rather (or also) the claim that only desire-dependent reasons can be reasons for action, which states a necessary condition. Searle does not dispute the sufficiency claim: it seems obvious that very often our actions can be adequately explained in terms of our needs and wants. It is the necessity claim, the idea that desires must be invoked in every adequate explanation of human action, which he challenges.

Searle claims not only that there are counter-examples to the Classical Model, i.e. cases of desire-independent reasons for action, but also that such cases are in fact abundant and can be found in many segments of our social lives. Searle describes the general mechanism operating in these cases, by means of which desire-independent reasons for action are generated, as follows. First, desire-independent reasons are in fact created by free rational agents when they perform certain actions; second, these desire-independent reasons are brought about in virtue of the fact that the agents act with the intention that their actions should create such reasons. Searle introduces two kinds of cases in which desire-independent reasons for action are created in that way: speech acts, like making statements and making promises; and everyday situations where practical reason is involved.

Statement-making gives us the simplest illustration of desire-independent reasons. When one makes a statement one is thereby committed to speaking truthfully. This commitment is not an extra element, additional to the speech act; it is something that is internal to the institution of statement making: a statement is by definition a commitment to the truth of the proposition expressed. Lying, which involves believing correctly that the proposition expressed is false, and being wrong, which involves believing incorrectly that the proposition expressed is true, both presuppose such an internal link between making a statement and being committed to the truth of what it expresses. Therefore, when one makes a statement one has thereby created a reason for oneself to tell the truth. Such a reason does not depend on any pre-existing desires of the agent; in particular, it is independent of any desire to tell the truth, or of any moral preference for the truth or inclination to speak truthfully.

Similar remarks hold with respect to promising, which is discussed at length by Searle. When one makes a promise one is thereby under the obligation to keep it. The obligation is not an extra element, additional to the speech act; it is something that is internal to the institution of promising: promises are by definition creations of obligations. Making insincere promises, promises made with no intention to keep, presupposes such an internal link between making a promise and being under the obligation to keep it. Therefore, given that obligations are reasons, when one makes a promise one has thereby created a reason for oneself to keep the promise. Such a reason does not depend on any desires one might have; in particular, it is independent of any desire one might have to keep the promise, or of any endorsement of a moral principle to the effect that one ought to keep one's promises. Searle discusses three common mistaken ideas about the source of the obligation to keep a promise: the idea that the obligation is prudential, the idea that it derives from our acceptance of moral rules such as the above principle, and the idea that if the obligation derives from the institution of promising, then nothing would prevent any other social institutions, including morally condemnable institutions like slavery, from creating similar obligations. These ideas stem from the Classical Model and from the divorce it is forced to postulate between the act of promising and the obligation to keep a promise. Searle rebuts the third idea by rejecting the premiss that the obligation to keep a promise derives from the institution of promising. Rather, the obligation is entirely created by the agent: when she performs an act of promising she freely and voluntarily creates a reason for herself to act in a certain way (i.e. to keep the promise made). Freedom of the will, which is absent in the case of slavery, is thus essential to the creation of desire-independent reasons for action.

Searle extends these results to cases of practical reason, especially cases where the performance of an act on a given occasion rationally grounds the performance of an act on a later occasion. The mechanism operating in such cases is essentially the one mentioned above. By performing now an act with the intention to create a desire-independent reason to perform a certain act in the future, a free rational agent now creates a desire-independent reason for herself to perform such an act in the future; hence, one can say that the agent now has a desire-independent reason to perform the act in question in the future. For instance, by ordering a beer now and drinking it one has intentionally created a commitment or obligation, and hence a reason, to pay for the beer later on; and such a reason does not depend on any desire one might have on the later occasion to pay for the beer. Searle offers a general theory of the role of desire-independent reasons in practical rationality, a theory in which the following five elements are combined: temporality, freedom, the first-person point of view, language and other institutional structures, and rationality. Temporality comes in because, in practical reason, we organise time in the sense of taking actions we perform now as causally responsible for actions we are supposed to perform in the future, regardless of what we really want to do then. Language and other institutional structures are essential because, as in the case of making a promise, or ordering a beer, there are crucial features of such structures that enable us to use them as tools in the creation of desire-independent reasons for future action. Freedom and the first-person perspective are essential because we freely and intentionally create reasons for future action when we act in a certain way in the present, on the one hand, and because the reasons so created are reasons for ourselves and not for anyone else, reasons we see as binding our own will in the future, on the other hand. Finally, rationality is essential in virtue of the requirements of logical consistency it imposes; in particular, "the agent should recognize the reason created as binding on his subsequent behavior" (p. 210).

We move now from action to another central relation between mind and world, knowledge, and more specifically to knowledge acquired by means of visual perception. In chapter 12 Zenon Pylyshyn employs the tools of cognitive psychology to address a set of important foundational issues in the theory of vision. These concern a crucial but often neglected aspect of the connection between our visual representations and the world, namely the way in which particular elements in our visual representations relate to particular objects in the world. The central problem Pylyshyn discusses is an interesting special case of the general problem of intentionality: how the mind is able to "reach out" and represent items in the world.

The problem is this. Consider a visual representation of a scene consisting of a number of particular objects, for example lines, or squares. On the plausible assumption that visual percepts are incrementally constructed by the visual system, two twin questions have then to be addressed. How can a particular part or token of the visual representation—as taken at a certain time or stage of construction—refer to, or pick out, a specific object in the visual scene? How can a token of the representation—as taken at a later time or stage—refer to, or pick out, the very same specific object? More crudely stated, the questions are these. How can a visual representation identify a specific object among a number of objects in the visual field? How can that very object be tracked or re-identified by the visual system as time passes and the representation is being built up? Such questions must be answered because the building up of a representation of a visual scene invariably involves the detection of new properties of the scene; so we must know to which part of the existing visual representation the freshly acquired information should be attached, and that depends on our being able to identify the specific object the information is about. Suppose that information of the form $P(a)$, where P is a property (e.g. redness) and a a specific object in the visual field (e.g. a particular square), has been already stored as part of a visual representation of a scene. Suppose we want to update such information with respect to a newly detected property Q (e.g. being right-angled). Then we must be able to distinguish between cases in which the updated information has the form of a property conjunction, $P(a) \wedge Q(a)$, where Q is predicated of the very same object, and cases in which it has the form $P(a) \wedge Q(b)$, where Q is predicated of a different object b in the visual field. In any case, the visual system must be able to identify or re-identify a specific object as the object the information encoded in a visual representation is about.

Pylyshyn contrasts two lines of explanation of how a reference to a specific object could be secured with respect to a given visual representation. (These lines of explanation have clear counterparts in accounts of linguistic and mental singular reference that are familiar from recent discussions in the areas of philosophy of language and philosophy of mind.) One is the descriptive model, a model according to which visual representations pick out specific objects by means of purely conceptual representations they supposedly provide, i.e. by means of sets of properties the objects uniquely exemplify. The vehicles or symbols that are appropriate to carry out such reference are *descriptors*, mental items that work like singular definite descriptions work in a natural language; thus, a visual representation picks out a specific object as long as this object happens to be the unique satisfier of the set of properties encoded in some descriptor. The other is the demonstrative model, a model according to which visual representations pick out specific objects in a direct manner, unmediated by any purely descriptive or conceptual representations. The vehicles or symbols that are appropriate to carry out such reference are what Pylyshyn calls *visual demonstratives* or *visual indexes*, mental items that work like names (or labels) and demonstratives work in a natural language; thus, a visual representation picks out a specific object as long as this object is demonstratively identified by some visual demonstrative, or is assigned to some visual index.

Pylyshyn's central thesis is that the reference mechanism at work in the case of visual representation is basically non-descriptive and demonstrative; it is the so-called FINST mechanism supplied by his theory of visual indexing (FINSTs, in other words 'fingers of instantiation', are visual indexes). This thesis is argued for not only on the basis of theoretical considerations but also on the basis of empirical evidence provided by a series of MOT (Multiple Object Tracking) studies conducted in Pylyshyn's laboratory.

A number of arguments are adduced against the descriptive model, the most important of which is that descriptive identification is hardly harmonizable with the incremental nature of visual representation. Indeed, if object identification by the visual system always proceeded via the employment of some descriptor, then the information-updating process with respect to a given object would consist in a continual updating of descriptors; it would thus involve a long series of retrievals of individuating descriptions and of the properties encoded therein. Hence, the descriptive model would demand amazing capacities of storage and retrieval, capacities we do not seem to exercise in very simple cases of continued reference to elements in the visual field. In contrast, the demonstrative model has the resources to account in a straightforward manner for the incremental elaboration of visual representations.

On Pylyshyn's view MOT studies provide evidence in support of the FINST theory of visual indexing. In MOT experiments subjects are asked to track a number of independently and unpredictably moving objects in a visual scene consisting of a set of qualitatively identical objects (e.g. circles); the experiments show that subjects are capable of tracking up to five such objects. Since the only way the moving objects are discernible from one another is via their locations, and since evidence shows that successful tracking does not proceed by storing locations and using descriptors based on spatial properties, it follows that visual elements are picked out and tracked directly, independent of any properties they have, including locations. According to Pylyshyn, these studies give credence to the hypothesis that there is in the early (pre-attentive) visual system a primitive mechanism that enables visual representations to pick out and track a small number of objects in a visual scene. This reference mechanism is primitive in the sense of pre-conceptual; in particular, it is prior to focal attention and property detection. Pylyshyn makes two important points concerning the question of how specific objects in a visual scene are linked to visual indexes (note that this is a one-to-one correlation, so that the number of visual indexes active in a visual representation is rather limited: it cannot be greater than five). On the one hand, specific objects are segregated from the background, separated from the rest of the visual field, in a primitive way, a way that is strongly similar to the familiar gestalt separation of figure from ground. On the other hand, visual indexes are attached to specific objects in the visual scene by virtue of data-driven causal chains that extend from these objects to certain symbol structures via primitive mechanisms of early vision; such causal connections constitute the element that ultimately assures that visual representations are able to pick out and track specific objects in the visual field.

The framework provided by the theory of visual indexing enables Pylyshyn to introduce a special category of objects, that of primitive visible objects. Primitive visible objects are by definition those objects that attract visual indexes and allow multiple object tracking. This notion of an object is contrasted with more usual notions, especially those according to which objects are individuated by means of sortal concepts. Moreover, unlike most objects in the usual sense, primitive visible objects are mind-dependent in the sense that they are identifiable in terms of certain features of our early visual system. Pylyshyn also notes some interesting ontological implications of his notion of a primitive visible object; the notion suggests an ontology where objects are assigned a primitive status and places a derivative one, where a conception of places as properties of objects replaces the usual conception of objects as properties of places.

The interrelations holding between epistemological questions, questions about the nature of our knowledge, metaphysical questions, questions about how the world is, and questions about the nature of our concepts and thoughts, are examined by Christopher Peacocke in chapter 11.

Concepts are the constituent units of intentional contents, the propositional contents of thoughts and other intentional mental states and of statements and other speech acts. Substantive theories of particular concepts are philosophical accounts of the nature and role of concepts employed in statements or thoughts belonging to specific domains, for instance mathematical thoughts or statements and modal thoughts or statements. Peacocke's central thesis is that there is a highly general task that substantive theories of particular concepts must carry out: in order to be adequate any such theory must satisfy a general condition that Peacocke calls the Integration Challenge. What a theory of particular concepts must integrate in order to meet the challenge are its metaphysical side, i.e. the account it gives of the truth-conditions of thoughts or statements in the particular domain covered by the theory, and its epistemological side, i.e. the account it gives of the methods by means of which such thoughts or statements can be known by us. Integrating the two aspects means providing an epistemology and a metaphysics for a given area of human thought that are not only separately acceptable but also coherent with one another in a certain specifiable way.

Besides containing a number of considerations meant to support the claim that the Integration Challenge exists and must be addressed by theories of particular concepts, Peacocke offers a detailed proposal concerning how the envisaged integration of metaphysics and epistemology should be in general attained and about what a substantive theory of particular concepts that meets the Integration Challenge would be like. Moreover, the way in which the challenge bears upon psychological theories of concepts is also discussed at length.

Why is the Integration Challenge a challenge? Why isn't the intended integration of metaphysics and epistemology a trivial task? The reason is that there happen to be various domains of thought and discourse with respect to which the tension between metaphysics and epistemology is conspicuous. It has been notoriously hard to reconcile an account of the truth-conditions of thoughts or statements in those areas with an account of their knowledge. Peacocke gives three paradigmatic examples of domains where the gap between metaphysics and epistemology seems difficult to close: modality, knowledge of the intentional contents of our own mental states, and the past. In some cases, for instance modality, the metaphysics is clear, at least if we assume—as we should on Peacocke's view—a realist account of the truth-conditions of modal statements; but the epistemology remains unclear for the truth of modal statements risks becoming, on that assumption, epistemically inaccessible. In other areas we have a clear epistemology but an unclear metaphysics; and there are still other areas of thought where neither is clear.

Peacocke goes on to devise a type of strategy that theories of particular concepts might follow with a view to meeting the Integration Challenge. His starting point is a substantive claim about concept individuation, a claim he calls the Linking Thesis. This thesis is in fact a conjunction of two claims: (1) there is a category of concepts, the epistemically individuated concepts, which are individuated by their role in the acquisition of knowledge, i.e. in terms of certain conditions under which intentional contents containing those concepts are known; (2) the category in question is fundamental in the sense that every concept either is itself an epistemically individuated concept or turns out to be partially individuated in terms of certain relations it bears to epistemically individuated concepts. The Linking Thesis establishes thus a tight connection between concepts and knowledge, between the identity of a concept and its epistemology, the way in which intentional contents containing it come to be known.

Peacocke mentions two kinds of epistemically individuated concepts: observational concepts like the concept of a red thing or surface; and logical concepts like the concept of conjunction. He also offers a general characterization of the notion along the following lines: a sufficient condition for a given concept to be an epistemically individuated concept is for that concept to be individuated in terms of its role in judgement, that is to say, in terms of certain conditions under which intentional contents containing the concept in question are judged or accepted by a thinker. This claim is supported by a detailed argument that is then used to unfold a way in which the Linking Thesis (together with a background supposition to the effect that a theory of particular concepts must determine an assignment of truth-conditions to contents containing the concepts) yields a general strategy to address the Integration Challenge. The strategy is applicable to a specific range of cases, namely those in which the following two conditions are satisfied: (*a*) a solution to the integration problem in the area relies fundamentally on truth-conditions; (*b*) the relevant concepts in the area are epistemically individuated concepts. The leading idea is roughly that it should be possible in such cases, via the Linking Thesis, to extract a determination of truth-conditions for intentional contents containing a target concept from an account of the role of the concept in judgement and hence in knowledge. Peacocke outlines two substantially different styles of solution to the integration problem, two ways of implementing the above strategy. One is the model of *constitutively causally sensitive conceptions*, which is shown to be adequate to account for the case of the past tense. The other is the model of *implicitly known principles*, which is shown to be adequate to account for the case of metaphysical necessity.

Peacocke closes with an extensive discussion of the psychological aspects of the Integration Challenge. He introduces two ways in which a psychological theory of understanding, in the sense of a characterization of a set of mental states and capacities required for the employment of concepts in a specific domain of thought, may fall short in providing a solution to the Integration Challenge. The first consists in what Peacocke designates as the insufficiency of the psychological theory and, on his view, is illustrated by Philip Johnson-Laird's theory about the nature of modal understanding. The second consists in the non-necessity of the psychological theory and is taken to be illustrated by Josep Perner and Ted Ruffman's theory of episodic memory. Peacocke argues that the difficulties afflicting these theories can be overcome only by devising more substantive theories of concepts, theories following the models mentioned above.

1 How Not to Find the Neural Correlate of Consciousness

Ned Block

The concept of H_2O and the concept of water are two concepts of the same thing, as are the concepts of temperature and mean molecular kinetic energy. But something like the converse case is also common. In the seventeenth century, there were two concepts whose referents did not coincide, both denoted by the ambiguous phrase now translated as 'degree of heat'. When they measured "degree of heat" by whether various sources could melt paraffin, heat source A came out hotter than B. When they measured "degree of heat" by how much ice a heat source could melt in a given time, B was hotter than A (Wiser and Carey 1983). We now know that the latter method measured heat whereas the former measured temperature. Aristotle used a term we translate as 'velocity' sometimes to mean average velocity, sometimes to mean instantaneous velocity. Again, his failure to see the distinction led to trouble.

Similarly, there are two concepts of "consciousness" that are easy to confuse with one another, *access-consciousness* and *phenomenal consciousness*. These two concepts of consciousness may or may not come to the same thing in the brain. The focus of this chapter is on the problems that arise when these two concepts of consciousness are conflated. I will argue that John Searle's reasoning about the function of consciousness goes wrong because he conflates the two senses. And Francis Crick and Christof Koch fall afoul of the ambiguity in arguing that visual area V1 is not part of the neural correlate of consciousness. Crick and Koch's work raises issues that suggest that these two concepts of consciousness may have different (though overlapping) neural correlates—despite Crick and Koch's implicit rejection of this idea.

I will start with two quotations from Searle. You will see what appears to be a contradiction, and I will later claim that the appearance of contradiction can be explained if one realizes that he is using two different concepts of consciousness. I'm not going to explain yet what the two concepts of consciousness are. That will come later, after I've presented Searle's apparent contradiction and Crick and Koch's surprising argument.

This chapter combines elements of a paper that appeared in *Trends in Neuroscience* 19/2 (1996) with other material, some from Block 1995. I am grateful to audiences at the 1996 consciousness conference in Tucson, at the 1996 cognitive science conference at the University of Siena, at the University of Oxford, Department of Experimental Psychology, at Union College Department of Philosophy and at the Royal Institute of Philosophy. I am also grateful to Susan Carey, Francis Crick, Martin Davies, Christof Koch, David Milner, and the editor of *Trends in Neuroscience* for comments on a previous draft.

Searle's Apparent Contradiction

Searle discusses my claim that there are two concepts of consciousness, arguing that I have confused modes of one kind with two different kinds:

There are lots of different degrees of consciousness, but door knobs, bits of chalk, and shingles are not conscious at all . . . These points, it seems to me, are misunderstood by Block. He refers to what he calls an "access sense of consciousness". On my account there is no such sense. I believe that he . . . [confuses] what I would call peripheral consciousness or inattentiveness with total unconsciousness. It is true, for example, that when I am driving my car "on automatic pilot" I am not paying much attention to the details of the road and the traffic. *But it is simply not true that I am totally unconscious of these phenomena. If I were, there would be a car crash.* We need therefore to make a distinction between the center of my attention, the focus of my consciousness on the one hand, and the periphery on the other. (Searle 1990: 632–4; italics added)

Note that Searle claims that if I became unconscious of the road while driving, the car would crash. Now compare the next argument.

the epileptic seizure rendered the patient totally unconscious, yet the patient continued to exhibit what would normally be called goal-directed behavior . . . In all these cases, we have complex forms of apparently goal-directed behavior without any consciousness. Now why could all behavior not be like that? Notice that in the cases, the patients were performing types of actions that were habitual, routine and memorized . . . normal, human, conscious behavior has a degree of flexibility and creativity that is absent from the Penfield cases of *the unconscious driver* and the unconscious pianist. Consciousness adds powers of discrimination and flexibility even to memorized routine activities . . . one of the evolutionary advantages conferred on us by consciousness is the much greater flexibility, sensitivity, and creativity we derive from being conscious. (Searle 1992)

Note that according to the first quotation, if I were to become unconscious (and therefore unconscious of the road and traffic), my car would crash. But in the second quotation, he accepts Penfield's description "totally unconscious" as applying to the case of the petit mal patient who drives home while having a seizure. Thus we have what looks like a contradiction.

Crick and Koch's Argument

I will now shift to Crick and Koch's recent article in *Nature* (Crick and Koch 1995*a*) arguing that V1 (the first major processing area for visual signals) is not part of the neural correlate of consciousness (what they call the NCC). Crick and Koch say that V1 is not part of the neural correlate of consciousness because V1 does not directly project to frontal cortex. (They extrapolate (tentatively) from the fact that no direct connections are known in macaques to no connections in humans.) Their reasoning makes use of the premise that part of the function of visual consciousness is to harness visual information in the service of the *direct* control of reasoning and decision-making that controls behavior. On the hypothesis that the frontal areas are involved in these mental functions, they argue that a necessary condition of inclusion in the NCC is direct projection to frontal areas. Though something seems right about their argument, it has nonetheless puzzled many readers. The puzzle is this: Why couldn't there be conscious activity in V1 despite its lack of direct connection to frontal cortex? This is Pollen's worry: "I see no a priori necessity for neurons in perceptual space to communicate directly with those in decision space" (Pollen 1995). The possibility of conscious activity in V1 is especially salient in the light of Crick and Koch's suggestion that visual consciousness is reverberatory activity in pyramidal cells of the lower layers of the visual cortex involving connections to the thalamus (Crick 1994). For one wonders how they have ruled out the possibility that such activity *exists* in V1 despite the lack of direct connection between V1 and frontal cortex. They do not address this possibility at all. The overall air of paradox is deepened by their claim that "Our hypothesis is thus rather subtle; if it [no direct connection] turns out to be true it [V1 is not part of the neural correlate of consciousness] will eventually come to be regarded as completely obvious" (Crick and Koch 1995a: 123). But the reader wonders why this is true at all, much less obviously true. When such accomplished researchers say such puzzling things, one has to wonder if one is understanding them properly.

I will argue that once the two concepts of consciousness are separated out, the argument turns out to be trivial on one reading and not clearly compelling on the other reading. That's the critical part of my comment on Crick and Koch, but I have two positive points as well. I argue that nonetheless their conclusion about V1 should be accepted, but for a different reason, one that they implicitly suggest and that deserves to be opened up to public scrutiny. Further, I argue that the considerations that they raise suggest that the two concepts of consciousness correspond to different neural correlates despite Crick and Koch's implicit rejection of this idea.

The Two Concepts

The two concepts of consciousness are *phenomenal* consciousness and *access-*consciousness (Block 1995). Phenomenal consciousness is just *experience*; access-consciousness is a kind of direct control. More exactly, a representation is access-conscious if it is actively poised for direct control of reasoning, reporting, and action.

One way to see the distinction between the two concepts is to consider the possibility of one without the other. Here is an illustration of access without phenomenal consciousness. In Anton's syndrome, blind patients do not realize that they are blind (though implicit knowledge of blindness can often be elicited). Hartmann *et al.* (1991) report a case of "reverse Anton's syndrome" in which the patient does not realize that he is not really blind. The patient regards himself as blind, and he is at chance at telling whether a room is illuminated or dark. But he has a small preserved island of V1 which allows him to read single words and recognize faces and facial expressions if they are presented to the upper right part of the visual field. When asked how he knows the word or the face, he says "it clicks" and denies that he sees the stimuli. There is no obvious factor in his social situation that would favor lying or self-deception. In addition to the damage in V1, he has bilateral parietal damage, including damage to the left inferior parietal lobe. Milner and Goodale (1995) have proposed that phenomenal consciousness requires ventral stream activity plus attention, and that the requisite attention can be blocked by parietal lesions. So perhaps this is a case of visual access without visual phenomenal consciousness. (Note that Milner and Goodale's account is not in conflict with Crick and Koch's claim that V1 is not part of the NCC if activity in V1 is not the object of attentional processes.)

So we see that access-consciousness without phenomenal consciousness makes sense and may even exist in a limited form. What about the converse, phenomenal consciousness without access? For an illustration at the conceptual level, consider the familiar phenomenon in which one notices that the refrigerator has just gone off. Sometimes one has the feeling that one has been hearing the noise all along, but without noticing it until it went off. One of the many possible explanations of what happens in such a case illustrates phenomenal consciousness without access-consciousness. Before the refrigerator went off, you had the experience (phenomenal consciousness of the noise, let us suppose) but there was insufficient attention directed towards it to allow direct control of speech, reasoning, or action. There might have been *indirect* control (the volume of your voice increased to compensate for the noise) but not direct control of the sort that happens when a representation is poised for free use as a premise in reasoning and can be freely reported. (It is this free use that characterizes access-consciousness.) On this hypothesis, there is a period in which one has phenomenal consciousness of the noise without access consciousness of it. Of course, there are alternative hypotheses, including more subtle ones in which there are degrees of access and degrees of phenomenality. One might have a moderate degree of both phenomenal consciousness of and access to the noise at first, then filters might reset the threshold for access, putting the stimulus below the threshold for direct control, until the refrigerator goes off and one notices the change. The degree of phenomenal consciousness and access consciousness may always match. Although phenomenal consciousness and access-consciousness differ conceptually (as do the concepts of water and H_2O), we don't know yet whether or not they really come to the same thing in the brain.

Once one sees the distinction, one sees many pure uses of both concepts. For example, the Freudian unconscious is *access*-unconscious. A repressed memory of torture in a red room could in principle be a phenomenally vivid image; what makes it unconscious in the Freudian sense is that it comes out in dreams, slips, fleeing from red rooms, and the like rather than directly controlling behavior. Thus in principle an image can be unconscious in one sense (not poised for access), yet experienced and therefore conscious in another sense (phenomenally).

Searle's Contradiction

Let's go back to Searle's (apparent) contradiction. You will recall that he says that if he were to become unconscious of the details of the road and traffic, the car would crash. "When I am driving my car 'on automatic pilot' I am not paying much attention to the details of the road and the traffic. But it is simply not true that I am totally unconscious of these phenomena. If I were, there would be a car crash." But he also says that Penfield's famous unconscious driver is "totally unconscious" yet manages to drive home. Note that there is no room for resolving the contradiction via appeal to the difference between 'conscious' and 'conscious of'. If Penfield's driver is "totally unconscious", then he is not conscious *of* anything. And thus we have a conflict with the idea that if one were to become unconscious of the road and traffic, the car would crash. Can we resolve the contradiction by supposing that what Searle thinks is that *normally* if one were to become unconscious of the road the car would crash, but the Penfield case is an abnormal exception? Not likely, since Searle's explicit conclusion is that consciousness adds flexibility, creativity, and sensitivity to action—suggesting that he thinks that consciousness is simply not necessary to routine activities like driving home. (Searle has never suggested that he changed his mind on this issue between 1990 and 1992.)

I think that appeal to the access/phenomenal distinction does serve to resolve the contradiction. The resolution is that Searle is presupposing that the Penfield petit mal seizure case loses phenomenal consciousness but still has sufficient access-consciousness to drive. But when he says that if he were unconscious of the road the car would crash, he is thinking of loss of both phenomenal and access-consciousness—and it is the loss of the latter that would make the car crash.

I find that audiences I have talked to about this issue tend to divide roughly evenly. Some use 'conscious' to mean phenomenal consciousness—to the extent that they control their uses. Others use 'conscious' to mean either access-consciousness or some kind of self-consciousness. But Searle's error shows how easy it is for people to mix the two concepts together, whatever their official stance is.

How Crick and Koch's Argument Depends on a Conflation

Crick and Koch argue that V1 is not part of the neural correlate of consciousness because V1 does not project to frontal cortex. Visual consciousness is used in harnessing visual information for directly guiding reasoning and decision-making and direct projection to frontal cortex is required for such a use. But what concept of consciousness are Crick and Koch deploying? They face a dilemma. If they mean phenomenal consciousness, then their argument is extremely interesting but unsound: their conclusion is *unjustified.* If they mean access-consciousness, their argument is *trivial.* Let me explain.

Let us look at their argument more closely. Here it is:

1. Neural machinery of visual consciousness harnesses visual information for *direct* control of reasoning and decision-making.
2. Frontal areas subserve these functions.
3. V1 does not project *directly* to frontal cortex.
4. *So* V1 is not part of neural correlate of consciousness.

Note that the 'direct' in premise 1 is necessary to generate the conclusion. But what reason is there to suppose that there cannot be *some* neural machinery of visual consciousness—V1, for example—that is part of the machinery of control of reasoning and decision-making, but only indirectly so? If by 'consciousness' we mean *phenomenal consciousness,* there is no such reason, and so premise 1 is unjustified. But suppose we take 'consciousness' to mean *access-consciousness.* Then premise 1 is *trivially* true. *Of course* the neural machinery of access-consciousness harnesses visual information for *direct* control since access consciousness just *is* direct control. But the trivial interpretation of premise 1 trivializes the argument. For to say that *if* V1 does not project directly to areas that control action, *then* V1 is not part of the neural correlate of *access*-consciousness is to say something that is very like the claim that *if* something is a sleeping pill, then it is dormitive. Once Crick and Koch tell us that V1 is not directly connected to centers of control, nothing is added by saying that V1 is not part of the neural correlate of consciousness in the *access* sense. For an access-conscious representation just *is* one that is poised for the direct control of reasoning and decision-making.

On this reading, we can understand Crick and Koch's remark about their thesis that "if it [V1 is not directly connected to centers of control] turns out to be true it [V1 is not part of the neural correlate of consciousness] will eventually come to be regarded as completely obvious." On the access-consciousness interpretation, this remark is like saying that if it turns out to be true that barbiturates cause sleep, their dormitivity will eventually come to be regarded as completely obvious.

To avoid misunderstanding, I must emphasize that I am not saying that it is a triviality that neurons in V1 are not directly connected to centers of control. That is an empirical claim, just as it is an empirical claim that barbiturates cause sleep. What is trivial is that if neurons in V1 are not directly connected to frontal areas, then neurons in V1 are not part of the neural correlate of access-consciousness. Similarly, it is trivial that if barbiturates cause sleep, then they are dormitive.

That was the "access-consciousness" interpretation. Now let us turn to the phenomenal interpretation. On this interpretation, their claim is very significant, but not obviously true. How do we know whether activity in V1 is phenomenally conscious without being access-conscious? As mentioned earlier, Crick and Koch's own hypothesis that phenomenal consciousness is reverberatory activity in the lower cortical layers makes this a real possibility. They can hardly rule out this consequence of their own view by fiat. Crick and Koch (1995*b*) say, "We know of no case in which a person has lost the whole prefrontal and premotor cortex, on both sides (including Broca's area), and can still see." But there are two concepts of seeing, just as there are two concepts of consciousness. If it is the phenomenal aspect of seeing that they are talking about, they are ignoring the real possibility that patients who have lost these frontal areas *can* see.

Crick and Koch attempt to justify the 'directly' by appeal to representations on the retina. These representations control but not directly; and they are not conscious either. Apparently, the idea is that if representations don't control directly, then they are not conscious. But this example cuts no ice. Retinal representations have *neither* phenomenal *nor* access-consciousness. So they do not address the issue of whether V1 representations might have phenomenal but not access-consciousness.

So Crick and Koch face a dilemma: their argument is either not substantive or not compelling.

Is the Point Verbal?

Crick and Koch often seem to have phenomenal consciousness in mind. For example, they orient themselves towards the problem of "a full accounting of the manner in which subjective experience arises from these cerebral processes...Why do we experience anything at all? What leads to a particular conscious experience (such as the blueness of blue)? Why are some aspects of subjective experience impossible to convey to other people (in other words, why are they private)?" (Crick and Koch 1995*c*).

Crick and Koch often use 'aware' and 'conscious' as synonyms, as does Crick in *The Astonishing Hypothesis*. For example, the thesis of the paper in *Nature* (Crick and Koch 1995*a*) is that V1 is not part of the neural correlate of consciousness and also that V1 is not part of the neural correlate of visual awareness. But sometimes they appear to use 'awareness' to *mean* access-consciousness. For example, "All we need to postulate is that, unless a visual area has a direct projection to at least one of [the frontal areas], the activities in that particular visual area will not enter visual *awareness* directly, because the activity of frontal areas is needed to allow a person to report *consciousness*" (1995*a*: 122; emphases added). What could 'consciousness' mean here? 'Consciousness' can't mean *access*-consciousness, since reporting is a kind of accessing, and there is no issue of *accessing* access-consciousness. Consciousness in the sense in which they mean it here is something that might conceivably exist even if it cannot be reported or otherwise accessed. And consciousness in this sense might exist in V1. Thus, when they implicitly acknowledge an access/phenomenal consciousness distinction, the possibility of phenomenal without access-consciousness looms large.

My point is not a verbal one. Whether we use 'consciousness' or 'phenomenal consciousness', 'awareness' or 'access-consciousness', the point is that there are two different concepts of the phenomenon or phenomena of interest. We have to acknowledge the possibility in principle that these two concepts pick out different phenomena. Two vs. one: that is not a verbal issue.

Are the Neural Correlates of the Two Kinds of Consciousness Different?

Perhaps there is evidence that the neural correlate of phenomenal consciousness is exactly the same as the neural correlate of access-consciousness. The idea that this is a conceptual difference without a real difference would make sense both of much of what Crick and Koch say and other researchers. But paradoxically, the idea that the neural correlates of the two concepts of consciousness coincide is one which Crick and Koch themselves actually give us reason to *reject*. Their original hypothesis—apparently about the neural correlate of visual *phenomenal* consciousness—is that it is localized in reverberatory circuits involving the thalamus and the lower layers of the visual cortex (Crick and Koch 1995*b*). This is a daring and controversial hypothesis. But it entails a much less daring and controversial conclusion: that the localization of visual phenomenal consciousness *does not involve the frontal cortex*. However, Crick and Koch now think that the neural correlate of consciousness does involve the frontal cortex, which would be a reasonable view about access-consciousness. Even if they are wrong about this, it would not be surprising if the brain areas involved in visual control of reasoning and reporting are not exactly the same as those involved in visual phenomenality.

One way for Crick and Koch to respond would be to include the neural correlates of *both* access- and phenomenal consciousness in the NCC. To see what is wrong with this, consider an analogy. The first sustained empirical investigation of heat phenomena was conducted by the Florentine experimenters in the seventeenth century. They didn't distinguish between temperature and heat, using a single word, roughly translatable as "degree of heat", for both. As mentioned earlier, this failure to make the distinction generated paradoxes. Recall that when they measured degree of heat by the test "Will it melt paraffin?" heat source A came out hotter than B, but when they measured degree of heat by how much ice a heat source could melt in a given time, B came out hotter than A (Wiser and Carey 1983). The concept of degree of heat was a *mongrel* concept, one that lumps together things that are very different (Block 1995).

The suggestion that the neural correlate of visual consciousness includes both the frontal lobes *and* the circuits involving the thalamus and the lower layers of the visual cortex would be like an advocate of the Florentine experimenters' concept of degree of heat saying that the molecular correlate of degree of heat includes both *mean* molecular kinetic energy (temperature) and *total* molecular kinetic energy (heat), which makes little sense. The right way to react to the discovery that a concept is a *mongrel* is to distinguish distinct tracks of scientific investigation corresponding to the distinct concepts, not to lump them together. No one would be in favor of using a concept of velocity for scientific purposes that lumped together instantaneous and average velocity.

Another way for Crick and Koch to react would be to include both the frontal lobes and the circuits involving the thalamus and the lower layers of the visual cortex in the neural correlate of *phenomenal* consciousness. (Koch seems inclined in this direction in correspondence.) But this would be like saying that the molecular correlate of *heat* includes both mean and total molecular kinetic energy. The criteria that Crick and Koch apply in localizing visual phenomenal consciousness are very fine-grained, allowing them to emphasize cortical layers 4, 5, and 6 in the visual areas. For example, they appeal to a difference in those layers between cats which are awake and cats which are in slow wave sleep, both exposed to the same visual stimuli. No doubt there are many differences between the sleeping and the waking cats in areas outside the visual cortex. But we would need a very good reason to include any of those other differences in the neural correlate of visual phenomenology as opposed, say, to the non-phenomenal cognitive processing of visual information.

A Better Reason for not Including V1 in the NCC

Though I find fault with one strand of Crick and Koch's reasoning about V1, I think there is another strand in the paper that does justify the conclusion, but for a reason that it would be good to have out in the open and to distinguish from the reasoning just discussed. (Koch tells me that what I say in this paragraph is close to what they had in mind.) They note that it is thought that representations in V1 do not exhibit the Land effect (color constancy).

But our experience, our phenomenal consciousness, does exhibit the Land effect, or so we would all judge. Similarly, it appears that neurons in V1 are sensitive to gratings that are finer than people judge they can make out. We should accept the methodological principle: *at this early stage of inquiry,* don't suppose that people are wildly wrong about their own experience. Following this principle and assuming that the claim that cells in V1 don't exhibit color constancy is confirmed, then we should accept for the moment that representations in V1 are not on the whole phenomenally conscious. This methodological principle is implicitly accepted throughout Crick and Koch's work.

An alternative route to the same conclusion would be the assumption that the neural correlate of phenomenal consciousness is "part of" the neural correlate of access-consciousness (and so there can be no phenomenal without access-consciousness). Phenomenal consciousness is automatically "broadcasted" in the brain, but perhaps there are other mechanisms of broadcasting. So even if the "reverse Anton's syndrome" case turns out to be access without phenomenal consciousness, Crick and Koch's conclusion might still stand.

Note that neither of the reasons given here makes any use of the finding that V1 is not directly connected to frontal areas.

The assumption that phenomenal consciousness is part of access consciousness is very empirically risky. One empirical phenomenon that favors taking phenomenal without access-consciousness seriously is the fact that phenomenal consciousness has a finer grain than access-consciousness based on memory representations. For example, normal people can recognize no more than eighty distinct pitches, but it appears that the number of distinct pitch-experiences is much greater. This is indicated (but not proven) by the fact that normal people can *discriminate* 1,400 different frequencies from one another (Raffman 1995). There are many more phenomenal experiences than there are concepts of them.

Despite these disagreements, I greatly admire Crick's and Koch's work on consciousness and have written a very positive review of Crick's book (Block 1996). Crick has written: "No longer need one spend time ... [enduring] the tedium of philosophers perpetually disagreeing with each other. Consciousness is now largely a scientific problem" (Crick 1996). I think this conceptual issue shows that, even if largely a scientific issue, it is not entirely one. There is still some value in a collaboration between philosophers and scientists on this topic.

2 Life and Cognition

Margaret Boden

1. Introduction

I must confess to feeling rather like the young man in the British film who, when warned by a suspicious policeman: "You want to watch it, sir!", anxiously replied: "Yes, officer—but where is it?" Any discussion of life and cognition must recognize that there is little agreement on what these terms mean.

Each is sometimes used in ways that presuppose the applicability of the other. I once heard a prominent A-Lifer define 'cognition' in terms of adaptation and responsiveness. But those criteria would include daisies and amoebae—which aren't obviously cognitive agents. That's not to say, of course, that they obviously aren't cognitive agents. Some philosophers, who define life as "autopoiesis", see all living things as cognitive systems: "The domain of all the interactions in which an autopoietic system can enter without loss of identity is its cognitive domain" (Maturana and Varela 1980: 119). Others would argue that cognition involves representation, and that it's because we don't ascribe representations to daisies and amoebae that we don't regard them as cognitive agents. Notoriously, however, the nature and even the existence of representations are now being challenged—and not only by proponents of autopoiesis (Clark 1997). But whether or not we see cognition as representation-based, it's not obvious that we should credit daisies with this capacity.

As for the reverse implication, many people seem to assume that intelligence presupposes life. This accounts for the popularity of the argument that since computers aren't alive (this also being taken as self-evident), they can't be intelligent. Perhaps this is right. But the dependence of intelligence on life, and the non-living nature of computers, should be explicitly argued, not taken for granted.

If intelligence does require life, it would follow not only that cognitive science is a sub-class of the life sciences, but also that AI—or at least, successful AI—is a sub-area of A-Life. If this conflicts with the history and the current sociology of these two fields, so much the worse for the researchers concerned. Or rather, so much the worse for (most) AI-researchers, given that A-Lifers are much more likely to see life as essential for intelligence. (This allows, of course, that AI may have something specific to contribute to A-Life as a whole, much as human psychology studies matters not covered in animal psychology.)

2. What is Life?

Although there is no universally agreed definition of life, a number of features are regularly mentioned in seeking one. They include self-organization, emergence, autonomy, growth, development, reproduction, adaptation, responsiveness, and metabolism. These seem to capture our everyday intuitions about the nature of life, and were listed long ago by Aristotle.

Some A-Life researchers hope to add newly discovered properties to this age-old list. For example, Christopher Langton (1991) has offered a quantitative expression of the common intuition, implicit in the catalogue just given, that life requires a subtle mix of change and stability. Langton defined a statistical measure of the order/disorder within a complex system, and calculated this number for an extensive range of computer models. He then watched them run, looking to see which ones produced performance which (like life) was neither wholly static, nor rigidly periodic, nor randomly chaotic. He found that interesting complexity arises only when his measure lies within a very narrow numerical range. In other words, life is possible—on Earth or elsewhere—only in a highly restricted, but precisely identifiable, range of systems.

If there is general agreement on anything in this controversial area, it would be the shared view that the central feature of life is self-organization. Self-organization involves the emergence (and maintenance) of order, out of an origin that is ordered to a lesser degree. It concerns not mere superficial change, but fundamental structural development. The development is spontaneous, or autonomous, in that it results from the intrinsic character of the system (often in interaction with the environment), rather than being imposed on it by some external force or designer.

To say that self-organization is central may be to make a weak claim or a stronger one. The weak claim is that it is the most important feature, the only one which is absolutely necessary for us to call something alive. The stronger claim is that self-organization potentially embraces all the others on the list. I favour the stronger claim. Emergence, autonomy, and development form part of the definition of self-organization given above. Metabolism and growth are aspects of self-maintenance, which also is explicitly mentioned. As for responsiveness, adaptation, and reproduction, I would argue, though shall not do so here, that they too are varieties of self-organization. Even on the stronger interpretation, however, self-organization does not suffice for life: it occurs in simple chemical systems (e.g. the Belousov–Zhabotinsky diffusion reaction). Life requires the more developed types of self-organization mentioned on the list.

One more item—evolution—is often added to the list of vital criteria. One might even argue that it is already implicit, for evolution assuredly fits the definition of self-organization given above. Evolution is commonly added because it appears to be the way in which terrestrial life has developed.

Moreover, it's the only scientifically intelligible process we know of which suffices to bring about this result.

That, at least, is the orthodox view. But some biologists and A-Lifers would object to the word 'suffice', insisting that evolution by natural selection is only one of the ways in which life has developed—and not the most fundamental of these. Brian Goodwin (1994) and Stuart Kauffman (1993) claim that the basic forms of life are originated by general principles of self-organization grounded in the physics of living matter. These (physical) principles are more fundamental than the (biological) self-organization of neo-Darwinian evolution. Natural selection merely fine-tunes the instantiation of these morphological forms, weeding out the organisms that are less well adapted to the relevant environment. Those who accept this unorthodox view can readily admit that evolution is a universal, and important, aspect of life's history on earth. They may even allow that it probably applies, as a matter of fact, to life elsewhere. But they may not list it as essential to life as such.

Even the biologically orthodox, however, should think twice before doing so. 'Thinking twice' is literally apt here, for there are two main reasons for not including evolution as a criterion of life.

On the one hand, to regard evolution as essential is to imply that the paradigm case of life is not an individual, but a group. Evolution is defined as a process occurring over successive generations, so cannot apply straightforwardly to living organisms as such. These can satisfy an evolutionary criterion only indirectly, as having originated within an evolutionary group. This flies in the face of our normal usage, which regards individual creatures as paradigms of life.

On the other hand, to say that evolution is essential implies that creationists are contradicting themselves when they claim that living things were specifically created by God. But this is too extreme. If creationism is nonsense (a charge I'd readily support), it is not non-sense. Or rather, if it is non-sense, that is not because its anti-evolutionary views are self-contradictory, but because some of its claims about the existence and/or nature of a creator God are non-sensical.

Nevertheless, the philosopher Mark Bedau (1996) argues that the crucial empirical role of evolution justifies our including it in the definition of life. He admits that this conflicts with our intuitions about the centrality of individuals and the conceivability of creationism. He admits, too, that it disallows our intuition that a non-evolving population—one which had (most improbably) reached a stable equilibrium—would be fully alive. But these difficulties, he says, should be ignored. It would not be the first time, after all, that an everyday concept has been modified by scientific advance. Compare the current and original meanings of 'sanguine' or 'phlegmatic', or consider Hilary Putnam's defence of adopting REM-sleep as a criterion of dreaming (Putnam 1962). Such conceptual modifications are only to be expected: ancient concepts will not always map neatly on to what modern science discovers to be natural kinds.

Non-philosophers may express impatience with this discussion, regarding it as a trivial terminological dispute. "If we allow that evolution is universal to life," they may say, "who cares whether we describe this as a defining criterion or as an exceptionless matter of fact?" If this dismissive remark means that it was a waste of time to discuss this matter at all, philosophers will demur: such discussions are essential in clarifying our concepts. But if it means that, after full discussion, we needn't worry about which words to use, many philosophers would agree. They might even quote Wittgenstein: "Say what you like!" I prefer to quote Wittgenstein's disciple John Wisdom: "Say what you like—but be careful!" Being careful, for Wisdom (as indeed for wisdom), involves not only exploring the reasons for saying the one thing and the reasons for saying the other, but also being aware of the likely implications for other discussions of favouring one usage rather than another.

In the present context, the "other discussions" concern the relation between life and cognition. One of our questions is "Do only living things have powers of cognition (or intelligence)—and if so, why?" The answer depends, in part, on the relation between evolution and meaning. If meaning is essential to cognition and if it can be grounded only in evolution (questions addressed in Section 3), then only evolved systems can be cognizers. So if we decide to include evolution as a criterion of life, we shall see life's potential to develop meaning as essential also. And this would make us more likely to believe that extra-terrestrial life, if it exists, probably includes some intelligent beings. The truth of the matter is unaffected by our definitions. But our expectations, and our aesthetic and religious responses, are not.

Whether only living things can be intelligent depends also on the extent to which evolution and the other listed criteria can occur in non-living systems. Perhaps some non-living, but evolved, artefacts might embody meaning? Evolution has already been simulated in various A-Life computer models, with some intriguingly lifelike results. To deny genuine life and/or genuine meaning to such systems, one must show either that the "evolution" involved is not genuine evolution or that some other essential criterion of life and/or meaning is missing. These matters are addressed in Section 4.

3. Cognition and Evolution

Cognition is the psychological category that covers perception, knowledge, language, and reasoning. As such, it exemplifies the vital criterion of responsiveness. Given that the cognitive apparatus possessed by each species picks up information that is relevant for its particular environmental niche, it is also a form of adaptation. In so far as this information aids survival, cognition contributes to self-maintenance (not as self-construction, but as self-protection). And it contributes to autonomy, too—especially in a language-using species, where it enables imaginative thought and deliberate action (Boden 1996).

What of the possibility raised in Section 2, that cognition may be essentially connected with evolution also? Here, we are not asking whether, as a matter of biological fact, cognition has emerged by means of evolution. Our question is rather whether it must do so, if it is to appear at all. Evolution would be necessary for cognition if it is necessary for meaning, which is a crucial—arguably, the crucial—cognitive phenomenon.

We cannot resolve this point by a simple equation of meaning with adaptiveness. The daisy's opening to the rising sun is adaptive, but not meaningful. This is not because no internal representations are involved: the cognitive role, if any, of representations is irrelevant here. Our reluctance to ascribe meaning to the daisy's autonomous movement lies rather in the fact that the daisy's response is so simple that there is no foothold for the threefold psychological categories of intentionality: cognition, motivation, and affect. All these are involved, if any one is, in meaningful behaviour. Indeed, all are involved in behaviour *tout court*, in so far as this is distinguished from mere movement or responsiveness. This was noted by the early critics of behaviourism, such as Ralph Perry (1921*a*, 1921*b*, 1921*c*) and Edward Tolman (1922), and is argued also in intentional systems theory (Dennett 1987). But if meaning differs from adaptiveness, it may nevertheless be adaptive in some philosophically interesting sense. If so, its relation to evolution could be significant.

Neo-Heideggerian philosophers such as Charles Taylor (1971) argue that meaning is inescapably elusive, since all ascriptions of meaning rest on interpretative assumptions which are themselves open to challenge. Meaning is seen as being rooted in human language and society, and is therefore denied to animals (and computers). Science, on this view, is an activity constructed by human subjectivity, and thus incapable of explaining it. No scientific explanation of meaning is possible, whether evolutionary or not.

I cannot disprove this constructivist philosophy, on pain of contradiction. But I find it highly implausible. The alternative is some form of realism, which will inform our account of meaning as well as our philosophy of science. If we are not to be trapped in ever-widening hermeneutic circles, and cast adrift by a fundamental anti-realism, there must be some way in which our ascriptions of meaning can be anchored in naturalistic phenomena.

A naturalistic account will ground meaning in some sort of causal commerce with the world. But what sort? Broadly speaking, such accounts fall into two types: the informational and the evolutionary.

Informational accounts define meaning in terms of reliable causal correlations between the environment and the behaviour and/or internal representations of the intentional system concerned. Such accounts are presented (for example) by Allen Newell and Herb Simon (1972: 21–6; Newell 1980).

Their definition of a physical symbol system takes designation and interpretation to be purely causal relations: "An expression [that is, a physical pattern in
the material system concerned] designates an object if, given the expression,
the system can either affect the object itself or behave in ways depending on
the object; and the system can interpret an expression if the expression
designates a process and if, given the expression, the system can carry out
the process." There is no requirement here that the material system involved be
made of flesh and blood, or that it have any particular individual or evolutionary history. It could be a newly manufactured computer, endowed
with cognitive capacities only ten seconds ago, on installation of a suitable
AI-program. Some informational semanticists would be less catholic in their
ascriptions of cognition. Fred Dretske (1988, 1994), for example, argues that
causal correlation is not the sole ground of meaning but must be supplemented by learning.

In the evolutionary approach, by contrast, the history of the species is
crucial. Causal correlations between the responses of living organisms
and their environmental circumstances are still regarded as essential, and
individual learning is allowed to strengthen the case for ascribing meaning.
But a certain type of evolutionary history is regarded as a *sine qua non* for
cognition.

The most well-developed evolutionary account of meaning is Ruth Millikan's (1984). Her account of propositional meaning, and of psychological
categories such as cognition and belief, rests on her concept of "normal
proper function" in biology. An organ's normal proper function (which
makes it what it is: a heart, for example) depends, among other things, on
the role that evolution has played in bringing about its current existence. A
spermatozoon's normal proper function is to fertilize an egg, even though it is
statistically *abnormal* for it to do so, since spermatazoa now exist only because
previous generations of spermatozoa contributed to the survival of the species
(and the generation of current individuals) by exercising their functional
capacity to fertilize eggs. Likewise, a heart's normal proper function is to
pump blood. An artificial heart, which has the functional capacity to pump
blood, is not really a heart, since its existence does not depend on a comparable evolutionary lineage. Evolutionary history, not merely current survival-
value, is essential to normal proper function—and to Millikan's account of
meaning and cognition, too. Broadly (the details are irrelevant here), meaning
is grounded in the adaptive roles that thought and language have played in
our evolutionary past.

It is not obvious that any naturalistic semantics is acceptable. Informational semantics in general faces difficulties in giving an account of error, and
other normative concepts. Since causal relationships are matters of empirical
fact, it is hard—many critics would say impossible—to see how they can
ground rational or linguistic norms. Evolutionary semanticists claim an
advantage here, arguing that norms can be grounded in the phenomenon of

adaptation. However, their non-naturalist opponents dispute this also (e.g. Morris 1992). Moreover, quite apart from the facts-to-norms difficulty, evolutionary theories of semantics face other problems which informational theories do not. For if we give an essential role to evolutionary history, we must resist ascriptions of meaning when that history is lacking—a requirement that can lead to highly counterintuitive conclusions.

Millikan offers a striking example, according to which an exact (atom-for-atom) simulacrum of a normal human being, behaving in all ways like the person it resembles, would be utterly devoid of understanding and intentionality. She imagines a cosmic accident whereby a host of atoms and molecules instantaneously coalesce to form a flesh-and-blood creature, of human appearance, in the middle of a swamp. Asked to name the capital of England, to predict the outcome of the next general election, or to evaluate the Mona Lisa, this creature says just what a suitably knowledgeable human might say. Asked to fill out an income-tax form, it sighs the same sighs and frowns at the same puzzling questions. But these utterances and gestures, according to Millikan, are meaning-less. Because evolution played no part in its history, the swamp-man has no beliefs about England, politics, taxes, or anything else. In short, this imaginary creature has no cognitive powers, despite its ability to mouth the expected sentences in response to our questions.

An informational semantics would not be embarrassed by this thought-experiment, for all the normal causal correlations are satisfied. And Millikan herself points out that, as the creature persists in its outwardly human life-style, which includes the capacity for learning, its causal links with the world would increasingly justify ascriptions of meaning based on a causal semantics. But this would be like classifying an artificial heart as a heart, because it fulfils the same biologically useful functions that a real heart does. Current causal links, even if they have a history of individual learning behind them, cannot satisfy an evolutionary criterion of meaning.

Millikan does not take this counterintuitive example as a knock-out blow. Much as Bedau is prepared to accept certain counterintuitive consequences of including evolution in his definition of life, so Millikan accepts such consequences in including evolution in her definition of cognition. She explicitly admits at the outset of her book that she cannot prove her philosophical account, or disprove others, as an a priori necessity. Rather, she presents it as a hypothesis to the best explanation. I share her view that an evolutionary semantics is the best naturalistic account of meaning available, while also agreeing (as noted above) that non-naturalistic theories cannot be disproved on pain of contradiction.

Evolutionary semantics, unlike purely informational accounts, explains meaning in a way that takes due account of theoretical biology. It treats cognition, including the propositional attitudes made possible by language, as part of our natural history. On this view, then, cognition is essentially connected with evolution—and with life. Indeed, the title of Millikan's book is *Language, Thought, and Other Biological Categories*, which makes it quite clear that, for her, life is essential for cognition.

Because evolutionary semanticists take it for granted that evolution is a specifically biological phenomenon, they rarely even discuss the question of whether meaning might properly be ascribed to artefacts. They regard the answer as obvious: computers, and other artefacts, cannot have cognitive powers, because they can neither evolve nor be alive. Section 4 asks whether this answer is correct.

4.   Could Artefacts Have Life or Cognition?

Artefacts do not occur naturally, but depend for their existence on human design. (Birds' nests and chimps' tools may be ignored here.) Accordingly, they appear to be paradigm cases of non-autonomous systems. Since autonomy is prominent in the list of vital criteria listed in Section 1, it therefore seems that artefacts cannot be alive. If no artefact is alive, and if life is essential for cognition, no artefact can be intelligent either. Moreover, one can argue against strong AI without making any explicit reference to life—as John Searle (1980*a*, 1992) does, for example. Many people would assume, then, that the two questions posed in this section have been answered: both life and cognition are foreign to artefacts.

However, things are not so simple. Life and meaning cannot be denied to artefacts so quickly. Searle's position is controversial, and in my view largely mistaken (Boden 1990). His argument that AI-programs lack intentionality rests on two claims: that their "meanings" are arbitrarily assigned to them by humans, and that neurochemistry (or some extraterrestrial equivalent) is necessary for cognition. The former claim, which relates to the discussion of Section 3, is addressed below. The latter claim is philosophically inexplicable (though conceivably true as a matter of fact), and irrelevant for our purposes. The crucial issue here is not biochemistry, but autonomy.

An autonomous system was defined in Section 2 as one whose behaviour results from its own intrinsic character (often in interaction with the environment), rather than being imposed on it by some external force or designer. Contrary to the common assumption cited above, some artefacts are autonomous. To be sure, they all (by definition) exist partly as a result of deliberate human action. Someone who focuses only on that baldly stated historical fact may insist that artefacts, being externally designed, have no intrinsic character—and therefore no autonomy. The possibility of autonomy resurfaces, however, if one considers the precise nature of the human intervention and/or the behaviour of the artificial system. Specifically, artefacts can be autonomous if they are evolved and/or embodied. (It does not follow that they are alive: autonomy is essential for life, but does not guarantee it.)

Before arguing this point, it is of interest to note that some such artefacts already exist. They range from creatures inhabiting virtual worlds, confined in computer memory, to material robots (designed or evolved) acting in and responding to a physical environment.

For example, the inhabitants of the software "Tierra" world are both evolved and virtual (Ray 1990). Tierran creatures appear on the computer screen as mere lines of code, but other creatures existing in virtual worlds include robots simulated by computer graphics, whose gross bodily anatomy and behavioural repertoire evolve over successive generations (Sims 1994). These simulated robots are "situated" in their (virtual) world: they are driven by the current environmental circumstances, not by some abstract internal plan. Some insect-like robots exemplify embodied, situated, artefacts produced by careful design (Brooks 1991). Other situated robots include physically embodied systems artificially generated by open-ended evolution (Cliff, Harvey, and Husbands 1993). The evolved features include not only the robot's behavioural strategies but also significant aspects of its sensorimotor anatomy and/or "neural" control-mechanisms. In cases of co-evolution, the task-environment itself evolves, offering challenges to later generations which earlier generations did not have to face (Ray 1990; Cliff and Miller 1995).

Evolution is relevant to autonomy because it generates things that are reasonably described as having their own intrinsic nature. Open-ended evo-lution is radically unpredictable. It is not restricted to exploring a given space of possibilities, for it can alter the dimensions of the current possibility-space (by allowing arbitrary changes to the length of the genome (Harvey 1992)). Since its end-results cannot be foreseen, they cannot be designed.

This applies not only to biological organisms, but also to artefacts. The end-results of artificial evolution would not have existed without the his-torical actions of the designer, who deliberately provided the evolutionary mechanism and the initial task-environment. But their current nature, which is discovered only by observation, is not due to design. Accordingly, it can be called intrinsic. (Admittedly, certain aspects will have been determined by the designer: whether the system is virtual or physical, for example, and what chemical composition, if any, it has. But these aspects were not subject to evolution.)

Non-evolutionary situated robots are often described as autonomous by their designers. This term applies to them only in a weaker sense. Their autonomy consists in the fact that their behaviour results relatively directly from their physical make-up, not indirectly from a computer program (Brooks 1991). Specific motor responses to specific environmental cues are built in. Some behaviour (often described as emergent) results from a com-bination of several direct responses and/or (directly cued) simple inhibitions of lower-level responses: a wall-following capability, for instance, may emerge which was neither built in nor predicted (Mataric 1991). A situated robot— whether evolved or designed—is autonomous in the sense that it is not a general-purpose machine, whose "nature" can be arbitrarily varied by human hand. It is what it is, and does what it does. As compared with a von Neumann computer, then, it might be said to have an intrinsic nature. But for non-evolved robots, the nature of this nature was carefully designed and specifically foreseen by the robot engineer. For evolved robots, that is not so.

In so far as evolution generates things with their own intrinsic natures, it is a source of autonomy—which is essential to life. Moreover, we saw in Section 2 that evolution is a form of self-organization, and that it is arguably essential to life, or at least a universal characteristic of it. Three core-features of life appear to be satisfied. So someone who denies that evolved artefacts are alive must show either that they did not really evolve, or that some other essential feature of life is missing.

The first line of attack rests on the fact that evolution is normally seen as a quintessentially biological concept. Someone might argue that A-Life systems, being non-biological, don't "really" evolve. This would be a mistake. Evolution is defined in terms of reproduction, heredity, variation, and selection. As such, it is an abstract informational concept whose most striking instantiation is biological (to say "most important" would raise questions about evolutionary cosmology). It is no accident that John von Neumann considered evolution in abstract computational terms, pointing out that it requires errors in self-replication—whose logical essence he defined prior to the discovery of the genetic code (Burks 1970). Nor is it surprising that evolution has been modelled by his successors, using the powerful computers that are now available. This modelling is still fairly primitive. Many aspects of genetic variation, natural selection, population genetics, and developmental epigenesis have not yet been simulated. But this is merely to say that current A-Life does not match the full richness of biology. It does not imply that evolution in A-Life is not real evolution.

What of the second line of attack, that some other aspect of life is missing? With one exception, all the items listed in Section 2 are (like evolution) abstract informational concepts, which in principle can be applied both to biological and to computational systems. The exception is metabolism, which refers not to information but to energy.

I do not see how this concept can be applied to virtual artefacts, or even to physical robots that do not maintain their bodies by self-regulated energy interchanges and chemical processes (Boden 1999). For metabolism is not the mere reliance on energy for fulfilling (vital) functions, but the use of networks of chemical analysis and synthesis for building and maintaining every part of the organism's bodily fabric. Mere battery recharging isn't enough, not even if its timing depends upon the robot's other activities and levels of "resource". Nor (*pace* Tom Ray 1994) is metabolism exemplified by simulated "energy-packets" attached to individuals within a virtual world. In short, A-Life's artefacts can't metabolize. This follows from the sorts of things that they (currently) are, not from the fact that they are artificial. Biochemical artefacts of more or less exotic types might metabolize (although there are empirical reasons for expecting that some sort of carbon chemistry will be involved). In short, if metabolism is essential to life, then (non-biochemical) artefacts can't be alive.

Whether metabolism is indeed essential to life might be seen as a "terminological" question of the type discussed in Section 2. If we can add evolution to the list of criteria despite Aristotle's omission of it, perhaps we can delete metabolism despite his—and everyone else's—inclusion of it? Metabolism would still be recognized as a universal characteristic of the sort of life we happen to know about, but it would no longer be seen as essential. This does not strike me as a sensible move. Whereas there are strong scientific reasons for opposing everyday intuitions about the non-criteriality of evolution, there are no good reasons for rejecting our intuitions about the necessity of metabolism. Moreover, to do so would arouse inappropriate fears in the wider community. The threat of self-replicating, and evolving, computer "viruses" is bad—and real—enough, without adding nightmares reminiscent of H. G. Wells. ("Say what you like—but be careful!")

If these remarks about the importance of metabolism are correct, then (computerized) artefacts can't be alive. Can they, however, have powers of cognition? Someone convinced that life is necessary for cognition must answer "No!" But is there any independent argument for denying cognition to artefacts?

One such argument has already been mentioned: Searle's claim that the "meanings" attributed to AI-programs are wholly derivative, being arbitrarily assigned to them by human beings. A given AI-program, he says, could in principle be used by us to stand for anything at all: a political process, a dance-routine, or a travel-itinerary. All that is required is that its syntactic structure maps onto the semantic structure of the relevant domain—whose meaning is available only to humans.

As a statement about traditional AI-programs, this is correct. But it's not clear (*pace* Searle's rejection of "The Robot Reply") that it applies to classical robots, which enter into genuine causal commerce with the world. And it's even less clear that it applies also to evolved, embodied, artefacts.

Evolution can ground intentionality, or meaning (see Section 3), and artificial evolution (I argued above) is real evolution. It follows that meanings could properly be ascribed to evolved robots whose behaviour, and even anatomy, has evolved so as to adapt to their task-environment—which itself may have evolved also. These meanings are neither arbitrary nor derivative. The justification for ascribing a certain meaning to the robot's behaviour at a certain time does not rest on syntactical matching, nor on foresighted design by a human being, but on its independently evolved, situated, behaviour. In short, such robotic meanings are intrinsically generated and environmentally grounded.

This claim may prompt resistance citing the weakness of current A-Life technology. The meanings one could ascribe to today's robots are certainly primitive, and arguably metaphorical. Current A-Life robots do not have advanced (models of) cognitive powers, nor of motivational and emotional mechanisms either. It follows that none of them embodies literal meaning, not even of the simplest kind. The point, however, is that today's (metaphorical) ascriptions of (primitive) meaning to an evolved embodied robot are not arbitrary, but are grounded in the robot's intrinsic nature.

Resistance might be based also on the difficulty of ascribing conscious intentionality to robots. Searle believes this difficulty is insuperable. Dan Dennett (1994) believes it is not. My own view is that to adopt either of these positions is premature (Boden 1998). Consciousness is so deeply problematic that the question must remain open.—"Please, officer, let me off: life and cognition are hard enough!"

3 The Representation of Number in Natural Language Syntax and in the Language of Thought: A Case Study of the Evolution and Development of Representational Resources

Susan Carey

> anthropology and psychology cannot be seen as truly independent disciplines. The centerpiece of anthropological theory is the centerpiece of psychological theory: a description of the reliably developing architecture of the human mind, a collection of cognitive adaptations. These evolved problem solvers are the engine that link mind, culture and world. Domain-specific performance is the signature of these evolved mechanisms, a signature that can lead us to a comprehensive mapping of the human mind.
>
> (Cosmides and Tooby 1994*b*: 111)

1. General Introduction

From an evolutionary perspective, it would be extremely surprising to find a nonhuman animal lacking in computational specializations. Simply take a mental walk through your average zoo or natural history museum, and consider bats and dolphins who use biosonar to auditorily see the world, honey bees that can detect polarized light, snakes that can see into the infrared, songbirds that can simultaneously tap into two independent sound production sources for singing their glorious courtship melodies, fish that can detect electric pulses, and both vertebrates and invertebrates that can generate their own light sources as advertisements for mating or defensive ploys against predators. These are indeed marvelous specializations and they illustrate the power of natural selection to fine tune an organism's neural machinery for computational tasks that allow it to solve species-typical ecological problems, or capitalize on species-typical ecological opportunities (Darwin 1859, 1871; Dawkins 1986).

This chapter is largely drawn from Hauser and Carey (1998). The research reported here was supported by NSF Grant SBR-97-12103 to M. Hauser and S. Carey.

Yet the causes and consequences of such variation in design features often go unnoticed in research on human cognition, and this is unfortunate. Just as in the case of nonhuman animals, it would be extremely surprising if it turned out that humans lacked unique, species-typical, computational abilities. One such specialization that we may be quite confident about underlies the human ability to learn, produce, and comprehend language (see synthetic pieces by Bickerton 1990; Chomsky 1986; Hauser 1996; Lieberman 1984; Pinker 1994). This capacity, however, is unlikely to be a single computational ability; presumably many different computational specializations serve language learning and use. Thinking about language in terms of species-typical adaptations raises research questions and provides research opportunities. Two related research questions immediately come to mind: what are the evolutionary roots of the human language capacity, and what are the relations among computational specializations for other cognitive tasks and those that underly language? Research opportunities derive from framing these questions in a comparative context.

The research program advocated here seeks to identify some of the basic cognitive building blocks for human language. Just as neuropsychologists search for functional dissociations as evidence for computational specializations, and developmental psychologists search for developmental dissociations as evidence for the building blocks of cognition, so studies of nonhuman populations are ideally suited for exploring the possibility of evolutionary dissociations. That is, in studying a diversity of species, one can address the hypothesis that a certain cognitive ability P could not evolve without the prior appearance of cognitive ability Q. More specifically, we can ask whether language could have evolved in the absence of particular cognitive capacities that are clearly a part of the human cognitive tool kit.

Evolutionarily-minded psychologists often claim that the time machine for relevant selection pressures on human cognitive algorithms need only be pushed back to the Plio-pleistocence period of hominid existence, or what has been called the "Environment of Evolutionary Adaptedness" or EEA (see papers in Barkow, Cosmides, and Tooby 1992). This general perspective is puzzling. As I demonstrate here, the roots of many important cognitive specializations are to be found early in primate evolution, if not before.

Except under conditions of relatively severe neural insult, all humans acquire one or more of the world's natural languages. What we do not know is which components of the learning mechanism supporting language acquisition are part of the species-specific learning device and which are computational devices that evolved prior to the evolution of the hominid branch of the phylogenetic tree. The research strategy is to look at certain key features that are expressed by human language (for example, as in the present case study, concepts of number and object, but one might equally examine concepts of agency, goal directedness, and intentionality) and ask whether they are necessary parts of the language acquisition device. Two sorts of evidence will ultimately bear on this problem: (1) are such abilities or representations available prelinguistically in human infants? (2) are such abilities or representations available to nonlinguistic primates or even, non-primates? We seek cognitive processes that babies or our primate cousins lack, as well as those they share in common with older humans. The end product of this research program, far in the future, would be an understanding of how the absence of particular computations prevents nonhuman animals and young human infants from acquiring a natural language, and how the presence of such computations facilitates language acquisition later in human life (Hauser 1996).

In order to make use of the comparative approach so familiar to biologists (Harvey and Pagel 1991; Ridley 1983), we must be sensitive to the fact that, in comparing species, similarity in the expression of a trait can arise from at least two different processes. On the one hand, two species can express the same trait because both have inherited the trait from a common ancestor. Here, we talk of the shared trait as a homology. For example, the five-digit hand that we see in humans and chimpanzees is a case of homology because the common ancestor of these two species expressed this particular morphological structure. In contrast, two distantly related species can express the same trait because individuals from each species confront a common ecological problem where selection has favored the same solution. This provides evidence of convergent evolution and here we talk of the shared trait as a homoplasy. For example, in songbirds and humans, young go through a period of vocal development where elements of the adult repertoire are produced in a primitive form, clearly influenced by auditory experience (Marler 1970; Locke 1993). This developmental trait, called babbling, is a homoplasy because songbirds and humans do not share a common ancestor that babbled. And yet the structure, function, and significance of babbling in these two distantly related taxonomic groups are similar.

The distinction between homology and homoplasy is important, for it provides us with important guidelines in selecting species for comparison. Moreover, it tells us that comparative studies require investigation into both taxonomically distal and proximal organisms. And the reason for such in-depth analyses into the tree of life is because we learn equally from cases where computational abilities differ between two species as from cases where they are similar.

If the comparative approach advocated is to succeed, we will require rigorous comparative methods. Studies of comparative cognition can be crudely divided into two categories. On the one hand are studies using highly controlled and sophisticated training techniques to examine, under captive conditions, what animals are capable of learning. Thus, for example, operant procedures have been used to explore the nature of nonhuman animal categorization processes and the structure of their conceptual representations (reviewed in Herrnstein 1991; Thompson 1995). Similarly, researchers working with apes and dolphins have taught individuals to use either formal sign language or an artificial language in an attempt to uncover some of the constraints on language acquisition (Premack 1986; Premack and Premack 1994*b*; reviewed in Roitblat *et al.* 1993). On the other hand are studies carried out under more naturalistic conditions, using a combination of observational and experimental techniques. For example, field and laboratory playback experiments have explored the putative meaning of nonhuman primate vocalizations, in addition to the cognitive processes underlying the implementation of such vocalizations (Cheney and Seyfarth 1990; Gouzoules, Gouzoules, and Marler 1984; Evans and Marler 1995; reviewed in Hauser 1996).

Both approaches have their merits (e.g. controls over stimulus presentation, ecologically relevant tests in the species-typical environment), but both use methods almost never used in studies of human cognition. This leads to the problem articulated by Macphail (1982, 1987, 1994), who claimed that comparative analyses were doomed to failure in the absence of comparative methods that could be used across species. The danger, Macphail warned, was that in using different methodological approaches one would be vulnerable to the criticism that inter-specific differences may be driven by method differences and inter-specific similarities may sometimes be spurious. This problem is most salient in animal/human comparative work, but the points apply in comparisons among nonhuman species as well.

Over the past four years, Marc Hauser and I have begun a comparative research program of human infant and nonhuman primate cognition that derives its power from the implementation of one experimental technique: the preferential looking time paradigm. The use of comparable methods to address cognitive processing across humans and nonhumans is, of course, not completely novel. There has been extensive work on animals from a Piagetian perspective, using the methods Piaget developed for studying sensorimotor development in human infants (see Antinucci 1989 for a review). Work by Adele Diamond (1988) on the neuro-cognitive constraints on object permanence tasks in human infants and rhesus macaques is an elegant testimony to what can be learned when the same methods are deployed across species. Both the Piagetian techniques and Diamond's work depend upon the reaching abilities of babies and primates, and are, we believe, somewhat more limited than the preferential looking time procedure with regard to the range of topics and species that can be explored.

I begin by briefly describing the general logic of the preferential looking time paradigm and then turn to a discussion of two broad classes of cognitive primitives that we are currently investigating from a comparative perspective. These two cognitive domains, concepts of objects and number, on the one hand, and agency, on the other, were selected because they appear to recruit non-linguistic concepts, but are fundamentally important to our capacity for linguistic processing. Thus, they seemed ideal domains to begin an exploration of the computational primitive subserving language from an evolutionary point of view.

A simple entry into the logic of the looking time procedure (see Spelke 1985 for a detailed discussion) is to consider what happens to us adults when we sit and watch a magic show. Consider a classic: the magician samples a random person from the audience, places him or her in an opaque box with feet emerging at one end and head emerging at the other end. The magician draws his trusty saw, shows the audience that it is for real, and then proceeds to saw the body in half. Having finished this seemingly gruesome task, he first separates the two halves, walks in between them, and then brings them back together. The presumably petrified audience member then walks out of the box, intact, and with no signs of blood or asymmetries in the way in which he or she was reassembled. And of course there are no signs of blood or body asymmetries. It was a trick. The body was never split in half. And yet, for all of our knowledge, we sit in awe, looking intensely, looking for the trick that violated our expectations. Violations of expectation are attention-grabbing, and so we can study patterns of attention to uncover the expectations non-verbal creatures have.

But our expectations as human adults need not be the same, or even similar, to those of the developing human infant, or more anciently evolved nonhuman. Armed with this logic, cognitive developmentalists have established the power of the preferential looking time technique in asking—sans language—what human infants know about the world, especially in its physical, mathematical, and psychological aspects (Leslie 1982, Leslie and Keeble 1987; Spelke 1991, 1994; Wynn 1992, 1996). In general, the technique starts with either a series of familiarization or habituation trials, designed to remove the effects of novelty on the subject's attention to the displayed events. Subsequently, the subject is presented with test events that are either consistent or inconsistent (i.e. from a normal adult's perspective; the terms "expected" and "unexpected", as well as "possible" and "impossible", are also used in a similar fashion) with the constraints provided by physical, mathematical, or psychological principles. Thus, for example, in order for a coffee cup to serve as a functional container for coffee, it must have a solid base, with no perforations in areas where the coffee is likely to rest. Knowing this, human adults would certainly be surprised (i.e. have their expectations violated) to see coffee remain inside a cup with a quarter-sized hole in its base. We would not, however, be surprised to see a 2 in. × 2 in. inch block of wood remain in the cup because it is larger than the hole, and is also a solid object. But is a human infant born with such expectations and, if not, what kinds of experience are critically involved in acquiring the relevant knowledge? And of equal interest, if such knowledge is innately specified, did we inherit it from our primate relatives?

2. Object and Number Primitives

In the psychological literature, the object and number concepts of nonlinguistic creatures are usually studied separately, often with distinct methodologies. Explorations of the infant's, or the animal's, concept of physical objects are often placed within a Piagetian framework (e.g. Antinucci 1989; Piaget 1955) or within the framework of studies of physical reasoning, which probe appreciation of contact causality, the solidity of objects, support relations among objects, etc. (e.g. Baillargeon 1994; Leslie 1994, Spelke 1991). Work on concepts of number, in contrast, focuses mainly on the representation of integers. The animal literature consists largely of conditioning studies which require discrimination of different numerical values (e.g. Capaldi 1993; Honig 1993; Meck and Church 1983), as well as a few studies of addition and subtraction (e.g. Boysen 1993), and of symbolic representation of integers (e.g. Boysen 1993; Matsuzawa 1985; Rumbaugh and Washburn 1993; Pepperberg 1994). The infant literature consists largely of studies of habituation to different numerical values (e.g. Antell and Keating 1983; Starkey and Cooper 1980; Strauss and Curtis 1981; Wynn 1996) and studies of arithmetical abilities which exploit the expectancy violation procedures described above (e.g. Koechlin, Dehaene, and Mehler 1996; Simon, Hespos, and Rochat 1995; Uller *et al.*, 1999; Wynn 1992, 1996).

Upon reflection, it seems clear that nonlinguistic object and number concepts must be studied together. Consider first our representation of integers; our procedures for enumeration. We cannot simply count what is in this room. Before we begin to count, we must decide what to count; we must individuate the entities in the room (where does one entity stop and another begin?) and we must know, when we look away and return to a portion of the room, which entities are the same ones as those we have counted before. That is, any enumeration procedure requires individuated entities as its input. The concepts which provide criteria for individuation and numerical identity (sameness in the sense of "same one") are called sortals in the philosophical literature, and in experiments in which objects are being counted, the questions arises as to the sortals under which they are being individuated. Consider next object permanence. All studies of object permanence are, in part, studies of criteria for numerical identity, for they involve the capacity to establish a representation of an individual, and trace this individual through time and through loss of perceptual contact. When we use the term "object permanence" to describe the baby's knowledge, we presuppose that he or she recognizes that the object retrieved from behind the barrier is the same one as the one that was hidden.

Thus, the exploration of the capacities of animals and infants to represent sortals is central to our understanding of the evolutionary and ontogenetic histories of both concepts: object and number. I will argue below (Section 2.4) that the connection is even closer, for what has been taken as evidence for counting by infants actually may not be, reflecting instead criteria for individuation and identity of physical objects, as well as a capacity to build short-term memory representations of small numbers of individual objects.

Here I explore the expression of number and of sortal concepts in human language, the acquisition of sortal concepts and numerical concepts in human infants, and present data from nonhuman primates that bear on the evolutionary history of the distinct primitives revealed in the human infant studies. I end with a speculation as to the evolutionary and ontogenetic sources of the uniquely human count system.

2.1. *The expression of number and sortal concepts in human language*

Not all languages have a counting sequence that expresses natural numbers. Nonetheless, numerical concepts are fundamental to all human languages. By "fundamental" I mean grammaticized; languages mark tens of thousands of conceptual contrasts lexically, and can express an uncountable number of ideas propositionally, but only a finite and relatively small set of concepts are reflected in grammatical contrasts. Number is universally reflected in grammaticized contrasts, the most important numerical primitive being the concept one. Number is typically grammatically marked on both nouns and verbs, usually reflecting the basic one/many (singular/plural) distinction, or sometimes reflecting three distinctions: one/two/many. In addition, noun quantifiers express numerical concepts: "an, another, few, many"; "an" picking out one, and "another" picking out a numerically distinct individual. "Few/many/some" all express subtly different contrasts from one. Finally, all languages have a grammatical particle that expresses numerical identity, sameness in the sense of same one.

"One" must be applied to an individuated entity. Thus, languages must represent concepts which pick out individuals, the sortals mentioned above. Sortals have been extensively studied in the philosophical literature on logic and semantics (Wiggins 1980; Hirsch 1982; see Macnamara 1986; Xu and Carey 1996, for discussion of sortal concepts within the context of psychological studies of concepts). In languages with a count/mass distinction, sortal concepts are expressed by count nouns, which is why they are called "count nouns", for they provide the criteria for individuation and numerical identity that enable entities to be counted.

Thus, number is reflected in language in three different ways, explicitly in counting sequences ("one, two, three..."), grammatically in number markers on nouns and verbs and the quantifier systems, and in the criteria for individuation and identity embodied in the sortal concepts the language lexicalizes. In the following sections, I ask which, if any, of the representational resources language makes use of in expressing numerical concepts are available to prelinguistic infants and nonhuman primates.

2.2.   Object as a primitive sortal, the quantifiers one, another: evidence from human infants

Piaget was the first to attempt to bring empirical data to bear on the question of whether human infants have a representation of objects as existing apart from themselves, apart from their own actions upon them and perceptual contact with them. In his famous studies of object permanence, he found that below around 8 months of age, babies will not retrieve an object that has been placed under or behind a barrier, and he interpreted this finding as revealing that they did not represent the object's continued existence. Indeed, Piaget argued that, even when babies succeed on this basic object permanence task, they do not yet represent objects as enduring through space/time, for they continue to display surprising failures on object search tasks, such as the A-not-B error. If babies of 8–11 months see an object disappear in location A, they will search in location A and retrieve it. Subsequently, if the object is then hidden in location B, the baby will search again in location A! Thus, Piaget concluded that the 8-month-olds' success reflected not an appreciation of object permanence, but rather a generalization: search where you see something disappear and something interesting will happen. Piaget argued that it was not until infants can reason about invisible displacements of objects (Stage 6 of the object permanence sequence; 18 months) that we can be certain that they represent object permanence. Piaget saw this achievement as part of the transition to symbolic thought, part of the transition to language.

Piaget's position differs from Spelke's nativist proposal that spatio-temporal continuity is one of the core principles that determines the entities in the world that actually are objects (a metaphysical claim). On Spelke's view, this property of objects does not have to be constructed over time; it is part of the set of innate principles that determines what infants take to be an object in the first place (an epistemological claim). Spelke offers data from the looking time methods described in the Introduction in support of her position. For example, Baillargeon, Spelke, and Wasserman (1985) and colleagues habituated babies to a screen rotating 180 degrees in front of the baby. It rotated toward and away from the infant. After habituation, a solid rectangular block was placed in the path of rotation of the screen, such that it was occluded from the baby's view as the screen rotated towards it. In possible events the screen stopped when it would hit the object behind the screen and then rotated back toward the infant (now traversing 135 degrees rather than 180 degrees). In impossible events, the screen rotated the full 180 degrees, apparently rotating through the object that was behind it. Babies as young as 3½ months remain habituated when viewing the possible events but dishabituate to the impossible events (Baillargeon 1987). This phenomenon shows:

1. Babies know that the object continues to exist behind the screen.

2. Babies know that one solid object cannot pass through the space occupied by another solid object.

Thus, besides demonstrating knowledge of object permanence, these data suggest that young infants also know that two objects cannot be in the same place at the same time (see also Spelke *et al.* 1992).

Many other experiments using the looking time methods support the conclusion that very young babies understand object permanence. Here I briefly describe two more, chosen because they illuminate the relation between object permanence and spatiotemporal criteria for individuation and numerical identity of objects, and because they also show that prelinguistic infants' representations are quantified by one and another. The first shows that babies do not merely expect objects to continue to exist through time, when out of view, but further, interpret apparent evidence for spatiotemporal discontinuity as evidence for two numerically distinct objects. Spelke *et al.* (1995) showed 4½-month-old babies two screens, from which objects emerged as in Figure 3.1. The objects were never visible together; their appearances were timed so that the movements would be consistent with a single object going back and forth behind the two screens. However, no object ever appeared in the space between the screens. Rather, one object emerged from the left edge of the left screen and then returned behind that screen, and, after a suitable delay, a second object emerged from the right edge of the right screen and then returned behind it. Babies were habituated to this event. Adults draw the inference that there must be two numerically distinct objects involved in this display, for objects trace spatiotemporally continuous paths— one object cannot get from point A to point B without tracing some continuous trajectory between the points. Spelke's babies made the same inference. If the screens were removed and only one object was revealed, they were surprised, as shown by longer looking at outcomes of one object than at the expected outcome of two objects. Control experiments established that infants were indeed analyzing the path of motion, and not, for example, expecting two objects just because there were two screens. That is, a different pattern of results obtains if an object appeared between the screens as it apparently went back and forth, emerging as before from either side. (See Xu and Carey 1996 for a replication with 10-month-olds.) These data show:

1. Infants know that objects continue to exist when they are invisible behind barriers.
2. Infants distinguish one object from two numerically distinct but physically similar objects (i.e. they have criteria for object individuation and numerical identity, and they distinguish one object from one object, another object.
3. Infants use spatiotemporal criteria for object individuation; if there is no spatiotemporally continuous path between successive appearances of what could be one or more than one object, they establish representations of at least two numerically distinct objects.

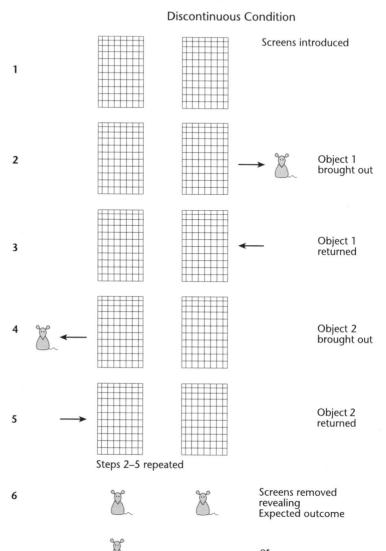

Figure 3.1 Schematic of Spelke, Kestenbaum, Simons, and Wein (1995) procedure.

Finally, Wynn's (1992, 1996) studies of infants' abilities to add and subtract provide further conclusive evidence that infants represent object permanence, and that spatiotemporal criteria determine object individuation and numerical identity. Wynn showed 4 ½-month-old infants an object, a Mickey Mouse doll, placed on a stage. She then occluded the doll from the infant's view by raising a screen, and introduced a second doll behind the screen, showing the infant an empty hand withdrawing from the screen. Then she lowered the screen, revealing either the possible outcome of two objects, or the impossible outcome of one object or of three objects. Infants looked longer at the unexpected outcomes of one object or three objects than at the expected outcome of two objects. Wynn also carried out a subtraction version of this study, beginning with two objects on the stage, occluding them with a screen, removing one from behind the screen, and upon lowering of the screen revealing either the possible outcome of one object or the impossible outcome of two objects. Again, 4½-month-olds looked longer at the unexpected outcome. Wynn interpreted these studies as showing that infants can add 1 + 1 to yield precisely 2, and that they can subtract 1 from 2 to yield 1. Before considering exactly how infants represent number (Section 2.4), here I emphasize the implications of Wynn's studies for infant representations of objects:

1. Infants represent objects as continuing to exist behind invisible barriers.
2. Infants distinguish two numerically distinct but physically similar dolls from one doll (i.e. infants have criteria for individuation and numerical identity: they distinguish one object from one object, another object).
3. Infants' criteria for individuation and numerical identity of objects are spatiotemporal, including principles such as one object cannot be in two places at the same time.

I conclude from these and other studies that, contrary to Piaget's position that the sortal concept, object, is built up slowly over the first two years of human life, it is most likely an innate primitive of the human conceptual system that serves to guide how experience shapes the development of physical knowledge. It is certainly available by 2 ½ to 3 months, way before it is expressed in natural languages. Also, the prelinguistic infants' representational resources include the basic quantifiers one, another. These two aspects of numerical representation, although central to human language, articulate infants' representations of the world prior to language production or comprehension. Of course, this provides only weak evidence that these quantificational devices did not evolve along with the computation resources that underlie the uniquely human linguistic capacity; they may be simply expressed earlier in development than other components of a uniquely human computational system for language. To address this question, I turn to animal studies.

2.3. *Object as a primitive sortal, the quantifiers one, another: evidence from nonhuman primates*

There is a long tradition of Piagetian studies of object permanence in non-human animals (dogs and cats: Doré and Dumas 1987; Gruber, Girgus, and Bannazizi 1971. Primates: see Antinucci 1989 for a review). Across a wide range of mammalian species, simple object permanence tasks are solved, and across a wide range of primate species, animals solve the A-not-B versions as well. That is, primates as evolutionarily distant from people as tufted capuchin and cebus monkeys and as evolutionarily close as gorillas and chimpanzees (Natale and Antinucci 1989) search for objects hidden under covers, and do not make the A-not-B error, performing as well as human infants of 15- to 18-months of age. There is also evidence that apes solve Stage 6 problems, although the success of monkeys at this level is more hotly contested (Natale and Antinucci 1989).

Primate success at Piagetian object search tasks is underscored by Adele Diamond's studies comparing rhesus and human infants on the delayed response task used in animals studies to study frontal function. Goldman-Rakic and Diamond noted the similarity between this task and Piaget's object permanence task (objects are hidden in one of two or more wells, a delay is introduced, and then the animal/baby is allowed to attempt to retrieve the object). Diamond showed that the development of success in delayed response in 2- to 4-month-old rhesus monkeys (including A-not-B errors, developmental increases in the delays that can be tolerated) is mirrored, in parametric detail, by parallel development in the same tasks in human infants between 7 and 9 months of age (see Diamond 1991 for a review).

Object permanence requires that a representation of an individual has been established, an individual that can be tracked through space and time. Given the evidence that prelinguistic infants as young as 4 months use spatiotemporal criteria for object individuation, we expected that primates from many branches of the evolutionary tree should also be found to rely on such criteria for individuation and identity in tasks such as those in Spelke *et al.* (1995) or Wynn (1992, 1996). Indeed, this result is anticipated by some of Tinklepaugh's observations in the 1920s (Tinklepaugh 1928, 1932). Tinklepaugh was one of the pioneers in the use of the delayed response methodology to explore object representations of monkeys, and he anticipated both the violation of expectancy paradigm and the Starkey (1992) use of number of reaches into a closed box as a reflection of numerical representations. On some trials in which the monkeys saw two pieces of food hidden in a given location, the experimenter surreptitiously removed one of them, so when the monkeys uncovered the food, they found only one piece. Tinkelpaugh reports that they were surprised, and that the number of reaches into the well was determined by the number of objects (one or two) the monkey saw placed there.

Before Hauser and I could pursue these observations more systematically, we faced a serious methodological challenge. Could we use the looking time methods with nonhuman primates? In the wild? In the laboratory? Would they be curious about the magic tricks we show to babies? Would they sit still and look at the outcomes? We saw these first studies as being as important for these methodological questions as for the scientific questions about criteria for individuation and numerical identity for objects.

Our first studies were of wild rhesus monkeys (*Macaca mulatta*) living on the island of Cayo Santiago, Puerto Rico, and we carried out three studies based on Wynn's addition and subtraction experiments (Hauser, MacNeilage, and Ware 1996). Our method was to catch the attention of a passing adult monkey, turn the video camera on him or her to record looking times, and show an event that either did or did not involve a magic trick. We made a box, open on the top, with removable screen that covered the front. There was a pouch on the inside surface of the screen, in which objects could be deposited on the trials with impossible outcomes, to be removed surreptitiously when the screen was removed. Our stimuli were bright purple eggplants, clearly food and thus of great interest to the monkeys, but unfamiliar food, so they were somewhat wary.

The first question we needed to answer was whether the monkeys would sit through the whole show, so our first shows were short. Unlike the infant methodology, in which looking times to both possible and impossible outcomes are recorded from each baby, often accompanied by within child baseline data, our first experiment with rhesus involved a single trial. Figure 3.2 shows the three trial types; twenty-four monkeys saw trials of type T1, in

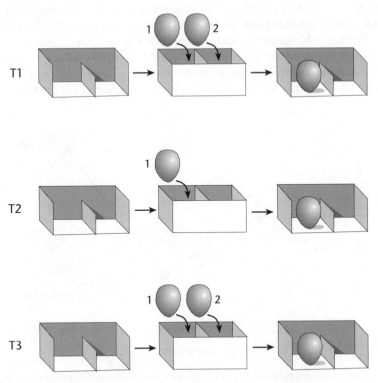

Figure 3.2 Schematic of Hauser, MacNeilage and Ware (1996) procedure.

which they were shown the empty box, the screen was put in place, and then one eggplant was withdrawn from the experimenter's apron, placed in the box, the empty hand shown the monkey, followed by a second eggplant's being withdrawn from the experimenter's apron and placed in the box. The screen was then removed, revealing the expected outcome of two eggplants in the box. Twenty-four monkeys watched trials of type T2, which were just like those of T1, except that only one eggplant was deposited in the box, and when the screen was removed, the expected outcome of one eggplant in the box was revealed. Both T1 and T2 were possible outcome trials. Finally, thirteen monkeys were shown trials of type T3, in which eggplants were placed into the box as in the T1 trials (i.e. 1 + 1), but the outcome was just a single eggplant, as in the T2 trials. Thus, T3 was the only trial type with an impossible outcome. Looking times were recorded for a fixed ten seconds after the screen was removed.

Like human babies, the monkeys looked longer at the impossible outcomes of the T3 trials than at the possible outcomes of either the T1 trials or the T2 trials. Thus, as expected, adult rhesus monkeys distinguish one object from two numerically distinct but physically similar objects, and they maintain representations of objects they see disappear behind barriers. These results were confirmed in a second series of experiments in which adult monkeys were shown to tolerate longer events. These experiments consisted of within-monkey comparisons of two baseline trials, followed by either T1, T2 or T3. In the baseline trials, no arithmetical operations were performed; they served to familiarize the monkeys with outcomes of both one and two eggplants in the box, but gave no information as to how many to expect in the test trials. Figure 3.3 shows the results. Monkeys looked markedly less on the second

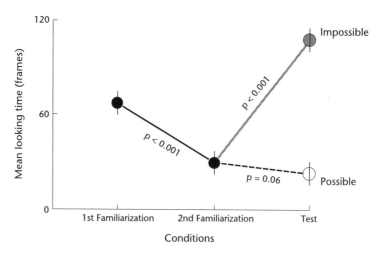

Figure 3.3 Looking times of Rhesus macaques in experiment schematized in Figure 3.2.

familiarization trial than on the first, and they maintained short looking times on each of the possible outcome trials (T1 and T2). On the impossible outcome trials (T3), monkeys looked markedly longer than on the possible outcome trials, and also markedly longer than on the second of the familiarization trials.

Thus, in both of these addition (1 + 1 = 2 or 1) studies, wild rhesus monkeys demonstrated surprise, in the form of long looking times at impossible outcomes. A third study in Hauser *et al.* (1996), compared looking times to two subtraction events (2 − 1 = 1, possible outcome; 2 − 1 = 2, impossible outcome.) Whereas the magnitude of the looking time differences was not as striking as in the two addition studies, the monkeys did look longer at the impossible outcomes than the possible outcomes, and also looked longer at the impossible outcomes than at the outcomes of the second of the two familiarization trials (although this difference was significant only on nonparametric measures.)

Methodologically, these studies establish that it is possible to use the looking time measures for studies of primate representations, even in the wild. Theoretically, they confirm the findings from Piagetian studies of object permanence that rhesus monkeys establish representations of objects, and track them through time, even out of view. They add to this literature the finding that, like human infants, rhesus monkeys distinguish between one object and two numerically distinct objects. They represent at least one sortal concept, *object*, and quantifiers such as one and another. These studies leave many questions open, of course, including the precise nature of the representations underlying performance on these tasks. The rest of Section 2 concerns the search for data that constrain our hypotheses about the nature of such representations in primates, including prelinguistic human babies.

Emboldened by these findings, we set out to establish whether these looking time procedures could be used to explore the representational systems of primates far more distant from human beings, evolutionarily, than are rhesus monkeys, namely small New World monkeys, cotton-top tamarins (*Saguinus oedipus oedipus*). These next studies posed new methodological challenges. Our sample was a colony of eight adult and two sub-adult tamarins, housed in the primate cognitive neuroscience laboratory of Harvard University, in four family groups. We needed to establish an experimental set-up in which we could carry out looking time experiments, and since there were only ten subjects, we could not afford to lose subjects due to inattention. With so few subjects, we could not use the between subjects designs of the studies of rhesus monkeys in the wild, and if we were to carry out a series of studies on related issues, we would have to establish that repeated testing with similar magic tricks did not lead to decreased interest in the relevant stimuli or classes of events. These studies are of theoretical interest, as well, as there have been few studies of object permanence with nonhuman primates more evolutionarily distant from humans than rhesus (see Antinucci 1989 for a review).

Our first experiments were four $1 + 1$ addition studies (Uller, Hauser, and Carey, forthcoming), two of which are briefly described here. Since tamarins are much smaller than rhesus, we constructed a smaller version of the box used in the studies described above, and rather than eggplants, used as our stimuli columns of the children's breakfast cereal known as fruit loops. Every few trials, after looking times were recorded, the monkey was given a fruit loop to eat. Intense interest in these treats helped maintain the monkey's cooperation through experimental sessions that consisted of four baseline trials (to establish a priori preferences for outcomes of one vs. two objects) and two test trials: $1 + 1 = 2$ (possible outcome) and $1 + 1 = 1$ (impossible outcome). The two test trials were identical to T1 and T3 of Figure 3.2.

Eight monkeys completed the experiment. Not surprisingly, given their interest in these stimuli, they had a baseline preference for displays of two objects over displays of one object. In sharp contrast, in the test trials the monkeys looked longer at the outcomes of one object (the impossible outcome) than at the outcomes of two objects (the possible outcome). The reversal of preference between baseline trials and test trials was significant on both parametric and nonparametric statistical measures.

These data extend the findings that rhesus monkeys look longer at the impossible outcomes of a $1 + 1 = 2$ or 1 addition task to cotton-top tamarins. Tamarins too individuate objects and track them through time, even out of sight. However, these studies, by themselves, do not show that rhesus and tamarin monkeys expect $1 + 1$ to be precisely 2. They are consistent with many other possible interpretations. For example, a possible interpretation of all of these studies that has never been ruled out holds that the monkeys (and the infants) are not tracking numerically distinct objects at all. Perhaps they are simply sensitive to amount of stuff, amount of fruit loop stuff in these studies, amount of eggplant in the wild rhesus studies, amount of Mickey Mouse stuff in Wynn's studies. Under this interpretation, the unexpected outcomes in the $1 + 1 = 2$ or 1 studies are unexpected because there is less stuff than the subject expected, not because there are fewer items.

Our second study tested this alternative hypothesis in a $1 + 1 = 2$ or big one study. That is, the outcomes of the test trials were either the expected outcome of two fruit loop columns or were a single large fruit loop column, twice the height of the single columns. If the monkeys are tracking amount of fruit loop stuff, or surface area of the stimuli, then both of these outcomes are possible and looking times should simply reflect whatever baseline preferences monkeys have between these two stimuli. However, if monkeys are tracking the number of objects, they should be surprised at the unexpected outcome of one larger object. The design for this study mirrored that for the previous study; there were four baseline trials that familiarized the monkeys with the two types of stimuli; two objects or one large object, being placed into the box, and established baseline preferences. No magic tricks were performed during the baseline trials. In the test trials, the objects were lowered, one at a time, into the box, as in T3 of Figure 3.2, and the outcomes were either two objects (possible outcome) or one large object (impossible outcome).

Nine of the ten monkeys completed all six trials. The baseline preference was for displays of two objects, and again, this preference was reversed in the test trials. The reversal was statistically reliable on both parametric and nonparametric tests. Apparently, monkeys are tracking individuated objects in these studies, not amount of stuff, or perceived surface area.

Just as did the studies with wild rhesus monkeys, these tamarin studies make several very important methodological points. The looking time methodology yields interpretable results with monkeys as evolutionarily distant from humans as are cotton-top tamarins. Further, it is possible to run a series of looking time experiments on the same small population of animals, with similar methods and materials, keeping 80–90 percent of the subjects on task through each of four experiments, and yielding interpretable findings throughout (see Uller, Hauser, and Carey, forthcoming, for a description of the whole series.)

These experiments show that cotton-top tamarins, like rhesus monkeys and preverbal human infants, represent the sortal *object*, with spatiotemporal criteria for individuation and numerical identity. In addition, the representational resources of these two species, like those of prelinguistic infants, include the basic quantifiers one, another. These two aspects of numerical representation, so central to human language, are part of the human primate heritage, and did not evolve along with the computational resources that underly the uniquely human linguistic capacity.

2.4. The representation of number by human infants

In Sections 2.2 and 2.3, we discussed the infant and primate addition/subtraction studies as they bore on nonlinguistic representations of object— object permanence, principles of individuation, and numerical identity for objects—and on nonlinguistic representations of basic quantifiers such as one, another. We argued that these are cognitive building blocks that emerged early in primate evolution, if not before. Here we return to the infant addition/subtraction studies as they bear on the question of prelinguistic infants' representation of number, especially the representation of the first three natural integers, 1, 2, 3.

Simple habituation experiments provide ample evidence that young infants, even neonates, are sensitive to numerical distinctions among sets of one, two, and three entities (e.g. dots: Antell and Keating 1983; sets of varied objects: Starkey and Cooper 1980; continuously moving figures: van Loosbroek and Smitsman 1990; jumps of a doll: Wynn 1996). In such studies, infants are habituated to arrays of a given set size (e.g. two entities), and then shown to dishabituate to arrays of a different set size (e.g. one or three entities.) Wynn's addition and subtraction studies confirm that prelinguistic infants discriminate among sets of 1, 2, and 3 objects, and additionally, that they know some of the numerical relations among them, for they have been shown to succeed at $1 + 1$, $2 - 1$, $2 + 1$, and $3 - 1$ tasks (Koechlin, Dehaene, and Mehler 1996; Simon, Hespos, and Rochat 1995; Uller *et al.* 1999; Wynn 1992, 1996).

The results presented so far leave open the nature of the representations underlying infants' performance. What these representations might be, and the senses in which they may or may not be "genuinely numerical" is a source of intense debate. In order to engage this debate, one must distinguish among classes of models that may underlie performance, and attempt to bring data to bear on which, if any, underlies infant performance. We know of three serious proposals for infant representation of number that could account for their successes in the studies cited above:

The Numeron List Proposal (Gelman and Gallistel 1978). Gelman and Gallistel proposed that infants establish numerical representations through a counting procedure that works as follows. There is an innate mentally represented list of symbols called "numerons": !, @, +, %, $ ... (of course, we do not know how such symbols would actually be written in the mind). Entities to be counted are put in one-to-one correspondence with items on this list, always proceeding in the same order through the list. The number of items in the set being counted is represented by the last item on the list reached, its numerical values determined by the ordinal position of that item in the list. For example, in the above list "@" represents the number 2, because "@" is the second item of the list.

The Accumulator Proposal (Meck and Church 1983). Meck and Church proposed that animals represent number with a magnitude that is an analog of number. The idea is simple—suppose that the nervous system has the equivalent of a pulse generator that generates activity at a constant rate, and a gate that can open to allow energy through to an accumulator that registers how much has been let through. When the animal is in counting mode, the gate is opened for a fixed amount of time (say 200 ms) for each item to be counted. The total energy accumulated will then be an analog representation of number. This system works as if length were used to represent number, e.g., "–" being a representation of 1, "—" a representation of 2, "——" a representation of 3, and so on (see Gallistel, 1990, for a summary of evidence for the accumulator model).

The Object File Proposal (Uller *et al.* 1999; Simon 1997). Babies may be establishing a mental model of the objects in the array. That is, they may be constructing an imagistic representation of the stage floor, the screen and the objects behind the screen, creating one object file (Kahneman and

Triesman 1984) for each object behind the screen. Such a model represents number, for example, the number 2, in virtue of being an instantiation of:

$$(\exists x)(\exists y)(\,(\text{object}(x)\&\text{object}(y))\&x \neq y\&\forall z((z = x) \vee (z = y))$$

In English this states that there is an entity and there is another entity numerically distinct from it, that each entity is an object, and there is no other object. This sentence is logically equivalent to "There are exactly two objects" but note that, in such a representation, there is no single symbol for the number 2 at all, not "2" or "@" or "—", or any other. This model exploits no representational resources other than those demonstrated in the previous sections: object sortals and the capacity to distinguish one from another.

Besides differing in the nature of the representation of integers, the three models differ in the process underlying discrepancy detection between the representation formed as objects are introduced (or removed from, in subtraction) behind the screen and the representation of the resultant display after the screen is removed. Take a $1 + 1 = 2$ or 1 event as an example. On the two symbolic models, the results of two counts are compared—the symbol for the number of objects resulting from the operations of adding (e.g. "@" or "—") is compared to the symbol resulting from a count of the objects in the outcome array ("@" or "—" in possible outcomes vs. "!" or "–" in impossible outcomes). According to the object model proposal, a representation consisting of two object-files constructed during the addition portion of the event is compared to a representation of two object files (possible outcome) or one object file (impossible outcome) by a process that detects 1–1 correspondence between the object files in the two representations.

These three proposals for nonlinguistic representational systems for number are genuinely different from each other. The first two (the numeron list model and the accumulator model) embody distinct symbols for each integer, but differ in the nature of the symbols they use. In the numeron list model each symbol bears a discrete and arbitrary relation to the number it represents. In the accumulator model, in contrast, an analog representational system exploits the fact that the symbols are magnitudes linearly related to the numbers they represent. And, as previously noted, in the object model system there is no distinct symbol that represents each integer at all. In this model there is nothing that corresponds to counting in terms of a set of symbols, whether arbitrary (numerons) or analog (states of the accumulator). Natural languages exploit the numeron list model; so we are particularly interested in evidence that this may be the representational system underlying the representation of number by prelinguistic infants and animals.

Uller *et al.* (1999) present several arguments in favor of the object file model as that which underlies performance on the infant addition and subtraction experiments (see also Simon 1997). The main argument is empirical: several experimental manipulations that might be expected to influence the robustness of mental models of the objects in the arrays, but not a symbolic representation of the number of individuals such as "@" or "—", are shown to affect performance of infants in the addition studies. To give just one example: the timing of the placement of the screen on the stage, relative to the placement of the objects behind it, determines success on a 1 + 1 = 2 or 1 addition study. The classic Wynn study (1992), and all the replications (Koechlin, Dehaene, and Mehler 1996; Simon, Hespos, and Rochat 1995; Wynn 1996) use an "object first" design (see Figure 3.3). The first object (1) is placed on the stage, then the screen is introduced, and then the second object (+ 1) is introduced behind the screen. Infants as young as 4 months of age succeed in this design. Uller *et al.* (1999) contrasted this design with a "screen first" design, in which the screen is placed on an empty stage, and then one object (1) is introduced behind it, and then a second (+ 1) is introduced behind it. Note, on the symbolic models, both of these designs simply require incrementing the counting mechanism twice, yielding a representation of two ("@" or "—"), and holding this symbol in memory until the screen is removed, so these two experimental designs should be equivalent in difficulty. But if we make some reasonable assumptions about the factors that might influence the robustness of mental models, then it seems likely that the object first design will be markedly easier than the screen first design. These assumptions are: (1) a mental model of an object actually seen on the stage is more robust than one constructed in imagery and (2) each update of a mental model in imagery decreases robustness of the model. The object first condition begins with a representation of one object on the stage constructed from perception and requires only one update in imagery; the screen first requires that the representation of the first object on the stage is constructed in imagery, and requires two updates in imagery. And indeed, infants succeed in object first tasks by 4–5 months of age, but in comparable screen first tasks not until 10 months of age (Uller *et al.* 1999). Note that in the rhesus and tamarin studies reviewed above the monkeys succeeded in the more difficult screen first design.

Other considerations favor the object file model as well, not the least of which is the finding of a sharp limit on the numerosities infants represent. Simple habituation experiments with infants, as well as the addition/subtraction studies, have shown that infants represent the numerical values of one, two, and three, but in general fail to discriminate among higher numerosities. There is no such limit on the accumulator model, or the numeron list model, but this limit is predicted by the object file model, on the assumption that there is a limit of parallel individuation of three object files in short term memory (see Trick and Pylyshyn 1994*b*).

In sum, I suggest that the weight of evidence currently available supports the proposal that the representation of number underlying infants' successes and failures in the addition/subtraction experiments, as well as habituation studies, consists of mental models of the objects in the arrays. These representations are numerical in that they require that the infant have criteria for numerical identity, because a representation that instantiates $(\exists x)(\exists y)(\,(\text{object}(x)\&\text{object}(y)\,)\&x \neq y\&\forall z(\,(z = x) \vee (z = y))$ is logically equivalent to "There are exactly two objects" and because comparisons among models are on the basis of 1–1 correspondences among object files. However, they fall short of symbolic representations of number, as there is no unique symbol for each integer, and because there is no counting process defined over them.

The upshot of this argument is that there is no evidence for a prelinguistic representational system of the same structure as natural language count sequences, such as "1, 2, 3, 4, 5 ...". There is no evidence from the infant studies that such a system is an antecedent available representational system, available to be exploited in the learning of language. The difficulty children experience learning to count (Wynn 1992, 1996) lends further credence to this conclusion.

2.5. The representation of number by nonhuman primates

There is a long history of studies of animal representation of number (see Boysen and Capaldi 1993 for a recent review of this large literature.) These studies have been of several types: some have required animals to learn to make responses contingent on some number of stimuli, or some number of previous responses; others have required the animal to choose the larger or smaller of two stimuli differing only in number of individuals. And recently there have been several studies of the ability of primates (mostly chimpanzees and rhesus monkeys; also one non-primate: an African grey parrot) to learn explicit symbols for integers, usually, Ø or 1 through 6, sometimes up through 8 or 9 (Matsuzawa 1985; Pepperberg 1994; Boysen 1993; Rumbaugh and Washburn 1993). All of these studies require extensive training, up to several years. The degree of training involved has led some commentators (e.g. Davis and Perusse 1988) to speculate that number is not a salient dimension of animals' experience of the world, and although they can be induced to encode number of objects or actions, they do not do so spontaneously. Others (e.g. Gallistel 1990) dispute this claim, arguing that number, at least in some ethologically relevant contexts such as foraging for food, is automatically encoded. The looking time studies with cotton-top tamarin and rhesus monkeys reviewed in Section 2.3 bear on this debate. Clearly, animals of both species automatically encode the difference between one vs. two entities, with no training whatsoever required.

The literature contains other demonstrations of nonhuman primate addition skills. Boysen (1993) reviews her studies of the chimpanzee Sheba, who had been previously trained to match arabic numerals (1–4) with numbers of objects. In the first addition study, Sheba went to one location, noted the number of oranges placed there (e.g. 2), went to a second location and noted the number of oranges placed there (e.g. 1), and then went to a third location and indicated the sum by pointing to one of four cards (1, 2, 3, and 4). The next studies dispensed with the oranges; Sheba went to the first location, looked at a card with an arabic numeral (e.g. 1), went to the second and looked at a card with an arabic numeral (e.g. 3), and then went to the third and pointed to the card which depicted the sum.

Rumbaugh and Washburn (1993) summarize another series of chimpanzee addition studies. Chimpanzees were presented with two sets of two trays of candies (e.g. 3 and 4 in one set and 2 and 6 in the other). The chimpanzee could choose one of the sets of trays, and was allowed to eat the candies from the selected set. The only incentive to add, or enumerate the total number of candies on one side, was to get the larger amount. The chimpanzees learned to do this. A variety of control experiments ruled out strategies such as taking the side with the largest single number of candies (e.g. the side with the tray with 6, which in the above example happened to be the correct choice), or avoiding the side with the smallest single number of candies.

Just as is the case for the infant studies, the important question in all the animal studies concerns the nature of the mental representations of number that subserve these abilities. In Section 2.4 I argued in favor of the position that representations of individual objects in the array underlie success on the infant addition and subtraction studies; that is, I favored the object file model. To date there are no relevant data that would bear on deciding among classes of models of the representations subserving the monkey $1 + 1 = 2$ or 1 studies. The object file model is a possibility, but an alternative is that the monkeys are spontaneously counting over some symbolic representation of number (as in the numeron model or the accumulator model).

It is important to note that some of the other primate addition studies are also consistent with the object file model. Boysen's original addition studies with Sheba are an example, assuming that 4 is within the limits of primate capacities for parallel individuation (see Trick and Pylyshyn 1994*b* for data that establish the human limit is as high as 4). Consider first the oranges version of the study. Upon encountering the first set of oranges, Sheba could have constructed a model containing an object file for each orange, and she could have updated this model by adding the number of object files that corresponded to the number of oranges she encountered second. To choose the correct arabic numeral, she would have had to learn to associate a representation with one open object file with "1", 2 open object files with "2", etc. This same association would allow her to solve the purely symbolic version of the task with the same strategy. That is, upon encountering the first numeral (e.g. "2") she would set up a memory representation with two open object files, and would update it by adding the number of object files depicted by the second numeral (e.g. "1"), yielding a model with a total of 3 object files open. She would then indicate the correct arabic numeral as in the version of the task that used oranges. Of course, it is also possible that Sheba's mental representations of 1, 2, 3, and 4 may have been symbolic, accumulator representations or numeron representations. The importance of the fact that Sheba had been taught external symbolic representations of number, that is, written numerals, will be discussed below. For now I wish only to stress that Sheba's achievements in these studies leave open the nature of the mental representations of number mapped onto those symbols.

Although the original addition experiments with Sheba are consistent with the object file model of numerical representation, it is also clear that primates must have additional resources for number representation. They must, because they succeed in representing numbers that exceed the limit on parallel individuation, which in humans is somewhere between 3 and 5. There is no reason to believe that nonhuman primates would have a greater capacity for parallel individuation than do humans. For example, in other experiments, Sheba learned arabic numerals up to 9, and also demonstrated knowledge of the ordinal relations among all pairs of numerals from 0 to 9. Squirrel monkeys, a new world primate, also have this latter ability; and in Rumbaugh's addition experiments, chimpanzees summed to 8 or 9 (Rumbaugh and Washburn 1993). These numbers exceed the memory limitations on concurrently opened object files, and implicate some other mental representational system for number.

I now turn to whether that system is likely to be the numeron list model or some analog system, such as the accumulator model suggested by Meck and Church (1983; see also Church and Broadbent 1990). Recall that the primary distinction made in Section 2.4 was that the numeron list model consists of a mentally represented list of arbitrary, discrete, symbols and a counting procedure that deploys them, whereas the analog model consists of magnitude representations and a procedure for establishing magnitudes that are proportional to the number of items to be enumerated.

The argument I would make concerning the nature of primate nonlinguistic representation of number is parallel to that made by Wynn (1992, 1996) concerning the nature of infant nonlinguistic representations of number. Wynn rejected the numeron list model for infant representation of number on the basis of an ease of learning argument. Note first that explicit linguistic representations, "1, 2, 3, 4, 5..." are a list of numerals. The only difference between the explicit numeral list representations of natural language and the posited nonlinguistic numeron list model is that numerons are mentally represented symbols, symbols in the language of thought. Wynn argued that if nonlinguistic number representation consists of a list of numerons, then learning to count in a natural language should be easy, and specifically, once children have identified the list in English (1, 2, 3, 4, etc.) that serves the purpose of the numeron list (!, @, +, %, etc.), they should demonstrate an understanding of how each of the English numerals functions in representing number. In her toddler counting experiments, Wynn showed that between ages 2½ and 3½ children can recite the count sequence, usually to 10, and can "count" in the sense that they can enumerate a set of items by tagging each one, reciting the count sequence in order, respecting 1–1 correspondence (see Gelman and Gallistel 1978, and Fuson 1988, for extensive studies of early counting). Importantly, Wynn also showed that children as young as 2½ know that "one" refers to the number one, that "two, three, four, five" all contrast with one and refer to numbers bigger than one. That is, they have identified the list as representing some aspects of number. Most crucially, Wynn showed that for a full year between ages 2½ and 3½ they don't know which number "2" refers to, or "3," or "4". Wynn argues that this state of affairs is inconsistent with the proposal that the underlying nonlinguistic representation of number is the numeron list model, for if it were, all that the child need do to understand how natural languages represent number is establish which list in natural languages functions as does the numeron list.

I read the literature on teaching symbolic representations of number to nonhuman primates as supporting a parallel argument. There are four series of attempts to teach primates symbols for number, all with chimpanzees (see Matsuzawa 1985, Boysen 1993, Rumbaugh and Washburn 1993, and Thomas 1992 for accounts of these experiments.) Each of these series of studies yields rich and fascinating information concerning chimpanzee representation of number; it is beyond the scope of this chapter to plumb all that is there. Here I wish to make three points. First, it is extremely difficult to teach chimpanzees symbols for numbers: Thomas (1992), 500,000 trials to learn binary symbols for 1 through 7; Matsusawa, 95 hours of continuous training to learn arabic numeral representations of 1–6 (compared to much less to learn a comparable number of object labels or color labels); Boysen, several years of daily training. Second, we are convinced that Boysen's chimpanzees, and Rumbaugh's, had learned to count, using a numeral list system. If the reader is not convinced that these studies have yet shown that, this just strengthens the point I wish to make here. Third, when one looks at the details of what is necessary to put this skill together, it is extremely unlikely that what the chimpanzees are doing is merely learning an explicit symbolic representational system that is structured identically to their nonlinguistic representational system for number.

Consider Boysen's studies as one example. Boysen taught Sheba, plus two other chimpanzees, Darrell and Kermit, a symbol system for number through several steps. First they were trained to make explicit 1–1 correspondences, to put one object into each one of six egg carton compartments. Then they were taught to match sets of one, two, or three objects with cards with one, two, or three dots. Then the cards with one, two, or three dots were replaced with cards with the symbols 1, 2, and 3. Numerals 4 and higher were taught directly as correspondences between the numeral and arrays of the relevant number of objects. Each new skill needed extensive training: after having learned to point to the correct numeral when shown a variety of arrays, new training was required for the chimpanzees to point to the correct array when shown a numeral, and still more to learn to construct arrays of 1, 2, 3, or 4 spools when shown one of the numerals (Sheba did not master this as well as Darrell, even though on most tasks she was the star pupil). More telling, even after two years of practice in these number numeral correspondences, two of the three chimpanzees failed to learn the serial order relations among the numerals 1–5 (Sheba succeeded; Darrell and Kermit failed). The other chimpanzees eventually succeeded, after another year of being trained to pick the smaller numeral as well as the larger one, and after further training on number/numeral correspondences, especially the construction task mentioned above. The end performance of these chimpanzees is very impressive indeed. Sheba spontaneously invented manual tagging (touching objects one at a time while enumerating them) to help in establishing number/numeral correspondences between arrays and numerals 4 or greater, and she eventually mastered the numerals up to 9, including the ordinal relations among them. That is, after being trained to pick the larger of two numerals from a subset of the numerals in her repertoire, she generalized to picking the larger from the pairs she'd never been trained on, and did so on the first exposure to them.

Two striking points emerge from these studies. First, the chimpanzees ultimately achieved significant mastery of an explicit numeral representational system for number. Second, given the extensive training required (not unlike preschool children, who also work at it for months before they can use a list of numerals to represent number), it appears that this is a new representational capacity being constructed, not simply an explicit expression of a numeron list structure antecedently available. Given that chimpanzees can learn explicit representations of numbers beyond the range of parallel individuation, it seems likely that one antecedently available representational system for number may be something like the analog accumulator model proposed by Meck and Church (1983).

In sum, I read the animal number literature as consistent with the claim that the numeron list system of representation of integers, widely but not universally expressed in natural languages, is a human cultural construction. It is not the representational system that underlies human infant appreciation of small numbers (here I favor the object file model), nor is it the representational system that underlies nonhuman primate representations of either small or larger numbers (here I favor the accumulator model, and leave open the possibility that the object file model is also available for the primates to draw upon in these tasks.) Of course, if analog representations of number are available widely in the animal kingdom, from rats through chimpanzees, it is very likely that these systems are part of the building blocks for human cognition as well. It is just that so far there is no evidence that human infants can exploit them.

2.6.   Specific object sortals: evidence from prelinguistic human infants

In Sections 2.2 and 2.3 I argued that prelinguistic infants and nonlinguistic primates represent at least one sortal concept, *object*, which provides spatio-temporal criteria for individuation and identity. But human adults use other types of information in establishing individuals, and tracing identity through time: property information and membership in kinds more specific than physical object. An example of use of property information: if we see a large red cup on a window sill, and later a small green cup there, we infer that two numerically distinct cups were on the sill, even though we have no spatio-temporal evidence to that effect (i.e. we didn't see both at once in different locations). With respect to information about kinds: adult individuation and numerical identity depends upon sortals more specific than physical object (Hirsch 1982; Macnamara 1986; Wiggins 1980). When a person, Joe Schmoe, dies, Joe ceases to exist, even though Joe's body still exists, at least for a while. The sortal *person* provides the criteria for identity of the entity referred to by the name "Joe Schmoe"; the sortal *body* provides different criteria for identity.

Recent data suggest that young infants represent only the sortal *object* and no more specific sortals such as *book, bottle, car, person, dog, ball*... That is, they represent only spatiotemporal criteria for individuation and identity, and not criteria that specify more specific kinds. Consider the event depicted in Figure 3.4.

An adult witnessing a truck emerge from behind and then disappear back behind the screen and then witnessing an elephant emerge from behind and then disappear behind the screen would infer that there are at least two objects behind the screen: a truck and an elephant. That adult would make this inference in the absence of any spatiotemporal evidence for two distinct objects, not having seen two at once or any suggestion of a discontinuous path through space and time. Adults trace identity relative to sortals such as "truck" and "elephant" and know that trucks do not turn into elephants.

Xu and Carey (1996) have carried out four experiments based on this design, and found that 10-month-old infants are not surprised at the unexpected outcome of only one object, even when the objects involved are highly familiar objects such as bottles, balls, cups, and books. By 12 months of age, infants make the adult inference, showing surprise at the unexpected outcome of a single object.

Xu, Carey and Welch (1999) found convergent evidence for the emergence of sortals more specific than *object* between ages 10 and 12 months. They habituated infants to the display of Figure 3.5, which adults see as a duck standing on top of a car. That is, adults use the kind difference between a yellow rubber duck and a red metal car to parse this display into two distinct individual objects, even in the absence of spatiotemporal evidence of the two objects moving independently of each other. In the test trials, the hand

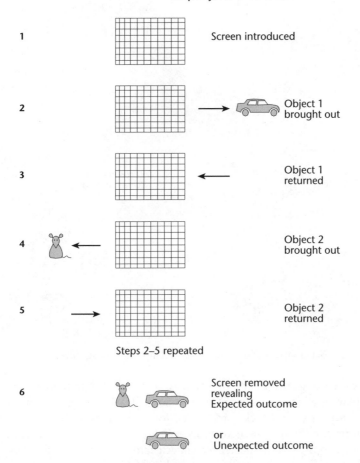

Property/Kind Condition

1   Screen introduced

2   Object 1
           brought out

3   Object 1
           returned

4   Object 2
           brought out

5   Object 2
           returned

Steps 2–5 repeated

6   Screen removed
           revealing
           Expected outcome

           or
           Unexpected outcome

Figure 3.4   Schematic of Xu and Carey (1996) procedure.

reached down and picked up the duck by its head; in the unexpected outcomes the single duck/car came up as a piece; in the expected outcomes, just
the duck was lifted by the hand. Ten-month-olds were not surprised at the
unexpected outcome; 12-month-olds, like adults, were, as revealed by longer
looking when the duck/car was raised as a single object.

Xu and I interpret these results as showing that before 12 months of age
infants use only the spatiotemporal criteria provided by the sortal *object* when
establishing representations of distinct objects in their mental models of the
world. By 12 months infants have constructed more specific sortals, such as
cup, bottle, car, ball, book, duck, and so on.

I am not claiming that young infants cannot represent properties of objects.
Indeed, very young infants can be habituated to different exemplars of

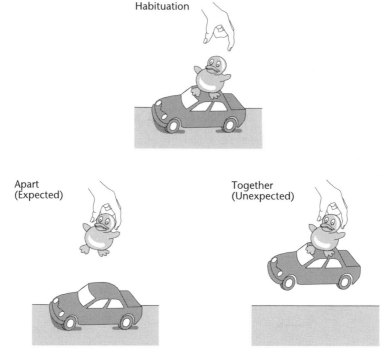

Figure 3.5   Schematic of Xu, Carey, and Welch (1999) procedure.

animals, or dogs, or tigers, or vehicles, and will dishabituate if shown an exemplar of a new category. Young infants clearly recognize bottles, cups, books, toy cars, toy ducks, and balls, for they know some object specific functions for them (which ones to roll, which ones to drink from, etc.). Similarly, young infants clearly recognize examples of person, for they expect people to move by themselves and to be able to causally interact without contact. And very young infants recognize particular people, such as their mothers. But none of these phenomena show that infants represent concepts like "a bottle, a book, a cup, Mama . . .", specific sortals or proper names that provide criteria of individuation and identity. One could recognize examples of objects which exemplify cuphood, or Mamaness, and have particular expectancies about objects with such properties, without representing Mama as a single enduring individual, or representing cup as a distinct sortal from book. Xu's results suggest that prior to age 12 months or so, such is the human infant's representational system.

It is significant that babies begin to comprehend and produce object names at about 10 to 12 months of age, the age at which they begin to use the differences between cups and elephants to individuate objects. In two different studies of highly familiar objects (bottle, ball, book, cup) Xu and Carey

(1996) found that comprehension of the words for these objects predicted the small number of 10-month-olds who could use these contrasts for object individuation. That is, babies do not seem to learn words for bottleshaped or bottleness; they begin to learn words such as "bottle" just when they show evidence for sortal concepts such as *bottle* that provide criteria for individuation and numerical identity.

These data raise a question that primate data could bear on. In human development, the construction of sortals more specific than *object* is contemporaneous with the earliest stages of language comprehension. Possibly, then, the capacity to represent concepts with the logical force of count nouns is part of the specific linguistic capacity of the human species. If so, we would not expect other primates to represent sortals for object kinds. Alternatively, the capacity to represent sortals more specific than *object* may be part of an evolutionarily ancient adaptation, way predating the emergence of human linguistic representational capacity, in which case we would expect other primates to behave as do 12-month-olds in tasks such as those of Xu and Carey.

2.7. Specific object sortals: evidence from nonhuman primates

The question at hand is not whether primates distinguish among categories of things—conspecifics from other animals, food from rocks—of course they do. The question is whether such categorical distinctions provide criteria for individuation and numerical identity of objects as the animal creates representations of the objects in its environment. That is, do primates, unlike young infants, take the difference between a cup and a tray to signal a difference between two numerically distinct individuals? At present, there are almost no data that bear on this question.

It is important to see what data do not decide this question, although they are suggestive. Tinklepaugh (1928, 1932) in the study cited in Section 2.3, included trials in which his monkeys saw the experimenter hide a piece of banana, and when they retrieved the food, found only a piece of lettuce. They searched for the banana, and showed great displeasure. Tinklepaugh concludes (and I agree) that their search was being guided by a representation of the actual object hidden. That is, they had set up a representation of an object with certain properties (banana properties) and they certainly failed to find a match to that representation, leading to disappointment. This observation shows that they can bind properties in an object file, once opened, but it does not tell us on what basis they open the file. It is important to note that the monkeys had spatiotemporal information that enabled them to open the object-with-banana-properties file; they saw the object placed in the container.

That they looked for the object-with-banana-properties when they found the lettuce suggests that they realized that the lettuce might be a different object from the one hidden, but not necessarily. When lettuce was hidden, and banana found, Tinklepaugh saw no evidence of surprise, or search for the lettuce; the monkeys happily ate the banana. Thus, the search for object-with-banana-properties may reflect extreme disappointment at not finding a match to the hidden object, the discovered lettuce being simply irrelevant.

What is needed is a design such as those of Xu and Carey, in which patterns of looking time establish how many objects the animal thinks are hidden behind a screen (Figure 3.4) or how many objects comprise an ambiguous display such as that of Figure 3.5, when only property or kind information licenses the animal to open two distinct object files. Together with colleagues, I have carried out one such study.

Free-ranging rhesus served as subjects (Uller *et al.* 1997). Twenty-four monkeys from the same population on Cayo as the studies described in Section 2.3 were run in a modified version of the Xu and Carey paradigm of Figure 3.4. The monkeys were shown an orange carrot emerging from one side of a screen and a yellow piece of squash emerging from the other; they never saw the two objects at once. After they were familiarized with these events, the screen was removed, revealing either the expected outcome of two objects (for half of the monkeys), or the unexpected outcome of one object (for the other half). Like 12-month-old human infants, the monkeys suc-ceeded on this task. They looked longer at the unexpected outcome of one object than at the expected outcome of two objects. They had used the differences between the carrot and the squash to establish representations of two numerically distinct objects.

These data, if confirmed in further studies that extend the range of object properties explored (e.g. which properties are bound to which objects and under which conditions), suggest that the capacity for individuation based on property or kind differences, as well as on the basis of spatiotemporal proper-ties, does not emerge only when the fully evolved human language faculty is present. At least one nonhuman primate species, evolutionarily more distant from humans than are the apes, also has this capacity.

2.8. Conclusions: ontogenetic and evolutionary building blocks for number representations

Early in primate evolution (and probably earlier) and early in the conceptual history of the child, several of the building blocks for a representation of number are firmly in place. Those discussed in this section include criteria for individuation and numerical identity (the sortal *object*; more specific sortals like *cup, carrot*), quantifiers such as *one* and *another*. Furthermore, there are conceptual abilities not dwelt upon, but which are equally important, such as the capacity to construct 1–1 correspondences and represent serial order relations (see Boysen 1993; Rumbaugh and Washburn 1993). Finally, there is no doubt that animals and babies are sensitive to numerical distinctions among sets of objects; that is, they represent number as one dimension of their experience of the world. These include representations of small numerosities (perhaps in the form of one, two, or three object files held in parallel in short term memory) and representations of larger numerosities (perhaps in the form of an analog representational system such as the accumulator model). All of these aspects of representations of number are prior, both evolutionarily and ontogenetically, to the linguistic expression of numerical concepts in the lexicon or syntax of natural languages.

We argued that the representation of the integers in terms of a list of numerals, or numerons (mentally represented numerals), is most likely a human cultural construction. Mastering it requires months, or years, of training, both by human children and by chimpanzees, suggesting that it is importantly different from the prelinguistic representations of numbers available to both infants and animals. The object file and accumulator models are importantly different from the numeron list model in just the required senses.

It is possible that this construction was made possible by human language. It is important to note that the process by which children master the numeral representational system for language differs from that by which chimpanzees have mastered those aspects of it they have (Boysen 1993; Rumbaugh and Washburn 1993). Human children learn the list of numerals, and the counting procedure, well before they map any of the numerals beyond 1 onto the numbers they represent. They then laboriously learn what "2" means, and then "3". By the time they have learned what "4" means, they have induced the principle by which the whole list represents number, and they immediately know what all the numbers in their count sequence mean (Wynn 1992). To date, no animal has been shown to make this induction, to be capable of learning an arbitrarily long list of numerons (many children can count to 10 or 20 before they know what "2" means). But although human language may be necessary for the original construction of a symbolic numeron representational system for number, and for the mastery of it through the process that normally developing children do so, chimpanzees at least can also master many important parts of the system.

In sum, I have so far found no evidence for any aspect of object/number representations that is part of the human specialization for language. Some I looked for clearly predate language capacities; others may require the language faculty but do not come for free to the child simply because of the human specialization for language learning.

4 The View from Here: The Nonsymbolic Structure of Spatial Representation

Ilya Farber, Will Peterman, and Patricia Smith Churchland

1. Introduction

One of the great challenges facing the cognitive and neural sciences right now is to figure out how nervous systems understand and represent space. Besides being intrinsically complex and fascinating, spatial representation is an important entry point to a range of large-scale questions of enduring philosophical interest. For starters, there is clearly a close relationship between the representation of space and the representation of one's own body, and if Antonio Damasio (1994) is correct body representation in turn underpins other forms of self-representation. Related questions involve the nature of coherence and unity in perceptual experience and motor control, and the processes whereby one modality can "trump" another for the sake of coherence. These lead to questions about modularity and how best to characterize such specialization as is seen in brains of mature animals, especially given interactive effects in brain development and in perception (Ballard 1991, Aloimonos 1993), and recent results concerning plasticity in the developing and adult brain (Elman *et al.* 1996; Finlay, Hersman and Darlington 1998). Finally, the study of spatial representation provides a fresh perspective on traditional debates about the nature and structure of mental representation, debates that have tended to focus exclusively on examples from the visual and propositional domains.

In this essay we will apply data from the neurobiology of spatial representation to one of the most enduring questions in cognitive science: is mental representation fundamentally symbolic?[1] Our view is that the psychology and neurobiology of spatial representation, to the degree that they are understood, do not sit well with a symbolic interpretation. In particular, recent results

---

[1] For the purposes of this essay, not much will turn on the precise definition of 'symbolic'. For clarity, however, let us stipulate that we understand the term to refer to systems which represent the state of the world in terms of the formal relationships among discrete logical tokens, and which represent changes in the world via syntactic manipulation of such tokens. Pure symbol systems, such as Turing machines and the formal languages of logic and mathematics, are those whose logical operations proceed without any reference to—or interference from—whatever physical structure they're implemented in. While PCs and natural languages lack this purity, symbolicists still regard them as *essentially* symbolic systems inasmuch as they can be compactly and profitably described as close approximations to ideal symbolic systems.

from neuroscience—at the cellular as well as the network and systems levels—suggest that the character of spatial representation is intimately dependent on the structure and physical organization of the underlying neural systems.[2] In contradiction to Steven Pinker's claim that "information and computation reside in patterns of data and in relations of logic that are *independent of the physical medium* that carries them" (1997: 34; emphasis added), the data surveyed below suggest that knowledge and reasoning about space seem highly *dependent* on the physical organization of the medium that carries them. If this is true, it presents proponents of the symbolic approach with some painful choices, which we will sharpen in our concluding discussion.

2.   Spatial Problem-Solving

Consider a contrast between what seems simply "given" and obvious, and what seems to require inferences or computation or "thought":

1. In a visual scene, one "directly" sees that the hammer is *on* the table, that the cat is *under* the bench, or that the flower is *to the right of* the tomato. Shifting one's gaze from the flower to the tomato is trivial; moving one's hand to grasp the flower or the tomato is trivial.
2. Children playing "hide and seek" in a park have to gauge which, amongst the available objects, will best screen their bodies from the view of the seeker. This is a nontrivial task, which will require time spent looking back and forth to judge lines of sight between the seeker and the various objects.

Philosophers schooled in a language-of-thought approach will tend to assume that only the second task involves mental representations and thought processes in any real or important sense, and that the processes enabling the two achievements are fundamentally different. In particular, it may be assumed that the second alone involves reasoning, reasoning which runs something like this: "I want to hide from Fred. He is going to see me unless I hide behind something. That bush is too short, that one is too narrow, that one is too sparse and transparent, so I will crouch behind this one."

We suggest that, appearances notwithstanding, the two kinds of operations are fundamentally related, and that understanding how the brain solves the "how should I hide myself?" problem will depend on understanding how the brain solves the problem of getting a representation of object-centered space. Both achievements depend on the brain's handling of spatial relations, undoubtedly a fundamental feature of brain organization in all animals. As

---

[2] Our essay thus connects to a tradition in psychology (Farah 1990; Stiles and Thal 1993; Kosslyn 1994) and linguistics (Elman and Zipser 1988; Kuhl 1991; Lakoff 1987) that seeks to explain at least some mental functions in terms of more embodiment-sensitive concepts such as mappings and dynamics.

we will show, the available evidence implies that the relevant representational systems are not language-like, and cannot readily be accounted for under a "symbol manipulation according to algorithms" paradigm.

Many animals routinely solve the hiding problem. Shirley Strum reports (in conversation) an example involving a female baboon who spies an out-troop male, potentially dangerous. When baboons are frightened, their tails reflexively rise up. In the example Strum reports, the female, partially shielded by a bush, reaches around to her back and pulls her tail back down out of the male's view.

Clark's nutcrackers hide nuts and, in later retrieval forays over the winter, remember which locations have been emptied and which still contain nuts. Grizzly bears have been observed hiding by ensuring that their considerable girth is well out of view of a human passing along a trail, though remaining quite visible to other creatures of lesser concern. This is not just the reflexive dive-for-a-dark-place behavior typical of a startled cockroach. It requires spatial understanding and knowledge of the relation of one's own body to other bodies in space.

Spectacularly, ravens can solve, sometimes in *one trial*, the problem of how to get a piece of meat hanging from twine tied to a trapeze (Heinrich 1993). The twine is set to a length that requires about seven iterations of this routine: "pull up a length with the beak, step on it, pull up another length with the beak, step on *it*, . . ." This task does not resemble any commonly performed by ravens in the wild, though ravens do step on a carcass to hold it down while ripping off chunks of meat. Crows also rip meat off a carcass in this way, but unlike ravens, they *cannot* solve the trapeze problem. The trapeze task requires the animal to perform a sequence of maneuvers, the first six of which have no immediate reward or sign of success except the increased proximity of the meat. This implies that the raven understands quite a lot about the spatial nature of the problem, and how its body needs to interact with the set-up to produce the right sort of change in spatial configuration. The point is, success in solving this problem is not achieved by trial and error, and it is not achieved by chaining in conditioning (in the way that pigeons can be trained, step by operant step, to perform a complex act). Success at the trapeze task relies on spatial representations.

One further example illustrates quite directly that spatial representation plays a role in food finding. Consider a rat put in a T maze where the left arm is always baited. The rat goes in at the bottom of the T, turns left, and gets the food (see Figure 4.1). Suppose now that a barrier is removed, converting the T maze to a cross maze, with the rat entering via the top arm. To get to the bait from this new starting point, the rat must turn right. If the rat has a genuinely spatial representation and understanding of its position in the maze, it will adjust for its new entrance path by turning right; if it has merely been conditioned to a "left turn" response, it will turn left as in the training condition. *Rats turn right.* Overtraining or hippocampal lesions can prevent

Training   Testing

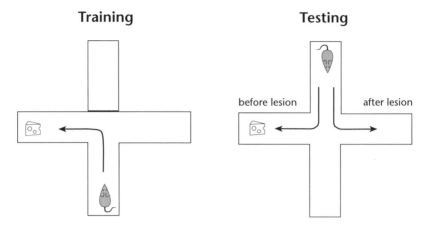

Figure 4.1 Spatial reasoning in the rat. In the training condition (left), the rat always starts in the same spot and learns to find food in another fixed spot. In the test condition (right), a block is removed and the rat is placed in the newly opened passage. Normally trained rats will turn right, correcting for their reversed spatial orientation with respect to the training location. Overtrained rats, or rats with hippocampal lesions, will turn left.

this compensation, causing the rats to turn left and miss the bait. (Packard and Teather 1998*a*, 1998*b*).

It is fairly safe to assume that the mechanisms baboons, bears, rats, ravens, and nutcrackers deploy in solving these tasks do not much resemble language-like propositional reasoning. Those who favor linguistic/symbolic models of human reasoning are thus left with a dilemma: either they must claim that humans and animals solve the same spatial problems in fundamentally different ways, or they must surrender spatial reasoning to the realm of the nonsymbolic and then search for some way to distinguish it in principle from other representational tasks of equivalent complexity. Neither path seems particularly promising.

For those (Fodor 1975, 1995; Pylyshyn 1984; Pinker 1997) who favor the "language of thought" hypothesis, symbols are valued because only symbols can be governed by formal rules, such as *modus ponens*. *Modus ponens* and its algorithmic kin don't work without formal symbols, and without such formal rules, the argument goes, we cannot explain problem solving, reasoning, and belief-desire structures generally. *With* formal rules and symbols, we can explain such profound things as the tendency of someone who believes "if P then Q" and "P" to also believe "Q".

We have no objection to the claim that trained humans occasionally perform *modus ponens*; but the very motivation for the symbolic project requires its proponents to claim much more. Two basic tenets of computationalism are that the fundamental architecture of thought is symbolic, and that normativity only makes sense in the context of symbolic representation (in other words, that you can't be *right* or *wrong* about something unless you're using symbols). Together, these principles gave rise to the standard methodology of "good old-fashioned" cognitive science: *start* with the basic resources of a symbolic computational system and then figure out how the system could be rigged up to exhibit the powers and properties exhibited by humans.

We find it more rewarding to look at the various problems involved in explaining mental processes from the other end. First, certain behavioral data imply that at least some forms of thinking and problem solving *are not* symbolic. Second, data from neuroscience and neural modeling reveal non-symbolic computational strategies whereby nervous systems can and do solve spatial problems such as reaching to a target, finding hidden food sources, and so on. Together, these findings provide a strong motivation for investigating the actual structure and dynamics of real neural systems—for *learning* from nature, rather than imposing our prior theoretical convictions on it.

Although considerable progress has been made in the last three or so decades, we still do not have anything like a comprehensive theory of the mind/brain, and there remain many unexplained functions: event storage and retrieval, attentional processes, cross-modal integration, pattern recognition, accurate prediction of events and their durations, flexibility in representation, the use of uncertain and incomplete knowledge, language learning, analogical inference, conscious awareness, skill learning, navigation, and so on *and on*. Nevertheless, such progress as *has* been made on these issues does not, in general, derive from the symbols-and-formal-rules paradigm. Hence Pinker's claim that "human thought and behavior, no matter how subtle and flexible, *could* be the product of a very complicated program" (1997: 37; emphasis added), however rousing a bit of boosterism for the symbolic approach, seems largely irrelevant to the project of understanding the way human thought and behavior *actually* work.

3.   The Neural Basis of Spatial Representation

In primitive animals such as the leech, spatial representations are essentially limited to location on the body surface. Receptors in the skin project to segmental ganglia which also house motor neurons that enable the leech to bend away from noxious stimuli. Anterior segments in the leech can also detect specific chemicals, and the leech can swim up a chemical gradient to a food source or down a chemical gradient to get away from something nasty. For these humble if nontrivial accomplishments, it need not have a representation of where, in external space and/or relative to itself, the targets are. In these sorts of animals, the spatial world and its body world are pretty much one. For the leech, "my body hurts here" is equivalent to "the world hurts here". As Damasio might put it, for such organisms, there is *only* subjectivity.

In fancier animals, evolution has stumbled on the strategy of exploiting a range of sensory signals from distinct modalities to get ever more accurate information, which can in turn support ever more sophisticated movement. Eyes, ears and vestibular systems, wings, and movable heads allowed animals to intercept moving prey detected at a distance, and sometimes even in the dark. Hearing and seeing are especially helpful in this regard, and the more so when auditory and visual signals *share constancies* such as intensity, source, and duration. But other signal-processing devices evolved too, such as electroreception (in fish) and infrared detection (in snakes), both of which register events at a distance. When there are independently movable parts, including movable signal detectors such as eyes, pinnae, whiskers, and antennae as well as movable heads and limbs, the brain has available a very rich story of "my body in the world". With mammals, and vertebrates in general, we see the emergence of a more complex, richer representation of the subject in an *objective* space—a space of things "out there".

What do we know about how neurons (in vertebrates) represent space? A vast literature is relevant here, but to keep within the limits of this essay we will restrict our focus to mammals. There are three areas in which substantial progress has been made in understanding how the mammalian brain represents objects in space. The first involves the hippocampus and has its roots in John O'Keefe's discovery in the 1970s of "place cells" in the hippocampus of the rat.[3] This line of research explores the role of cells in the hippocampus in representing and navigating through space. It turns out that a given place cell may code for a different region of extra-body space in different environments, and that the "maps" are not topographical. The findings so far suggest that these spatial coordinates are allocentric rather than egocentric (in other words, based on an external frame of reference rather than on a self-centered one). Some recent results indicate a relation between spatial learning, place cell responsivity, and rehearsal of place-finding in dreaming (Wilson and McNaughten 1994).

The second pertains to discoveries by Fuster and by Goldman-Rakic of cells in the prefrontal cortex of the monkey that hold spatial information in working memory. Other research (Fogassi *et al.* 1992, Colby and Duhamel 1993, Graziano and Gross 1993) revealed multi-modal cells in prefrontal cortex—visuo-somatosensory cells—which encode the position of objects in *head-centered* coordinates (see below). The third research line concerns structures in the posterior parietal cortex of the monkey (areas 5 and 7) that provide the substrate for transforming signals from retinotopic coordinates to body-centered or eye-centered or object-centered coordinates. (See Figure 4.2 for the locations of these structures in the macaque monkey.)

The three brain areas of particular interest here (hippocampus, prefrontal, and posterior parietal) are also highly interconnected, so it may not be foolish

---

[3] For a more recent account, see Wilson and McNaughten 1993.

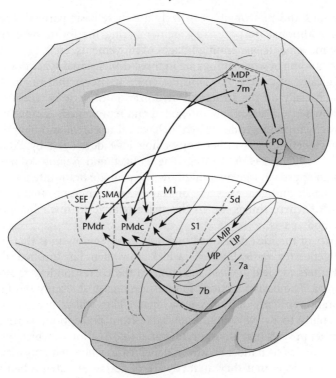

Figure 4.2 Connections to and from posterior parietal cortex, shown in medial (top) and lateral (bottom) views of the left hemisphere of a macaque. Arrows are shown as unidirectional, but most corticocortical projections are reciprocal. Areas included in the posterior parietal cortex (discussed in Section 3, below) are 5a, 7a, 7b and 7m, the medial dorsal parietal area (MDP), and the medial, lateral, and ventral intraparietal areas (MIP, LIP, VIP). Other abbreviations: P0, parieto-occipital visual area; SI, somatosensory cortex; Ml, primary motor cortex; PMd, dorsal premotor cortex; SMA, supplementary motor area; SEF, supplementary eye field. From Wise *et al.* (1997), used by permission of the authors.

to hope for a consilient, interlocking theory in the long run.[4] Research at the cellular level on prefrontal and posterior parietal has been done mainly on the macaque monkey, though human neuropsychological data suggest it is reasonable to expect a roughly comparable story for humans.

Our focus will be on the posterior parietal cortex. This area appears to provide the fundamental "objects out there external to my body" organization critical for primate sensorimotor representation and control. The

---

[4] Nor indeed are these the only areas that play a role in spatial knowledge. To do justice to the matter, one should also discuss the superior colliculus (Groh and Sparks 1996), ventral stream (Milner and Goodale 1995), and probably the cerebellum and the basal ganglia (see esp. Jeannerod 1988; and Milner and Goodale 1995).

hippocampus and prefrontal areas likely use these basic parietal representations for additional purposes (e.g. remembering the locations of goodies, planning movements, generating images of movements, etc.).

Understanding of where things are in three-dimensional space does not just arise magically, of course, nor is it just *given*, whatever that might mean. It depends crucially on the structural organization of various receptor sheets, and on how sensory signals are integrated and represented. A range of results from basic neurobiological research, behavioral research, and neural modeling come together in a rather compelling idea developed by Pouget and Sejnowski (1997*a*, 1997*b*, 1997*c*). The Pouget and Sejnowski hypothesis grounds an approach to explaining how the primate brain integrates diverse sensory signals and generates an "objective" representation, that is, a representation of where things are in the space relative to one's independently movable body parts.

The crux of the idea developed by Pouget and Sejnowski is that posterior parietal generates *basis functions* which can then be "pushed through" assorted filters to yield "go-to" locations in the corresponding motor reference frames. The basis functions in question are most simply thought of as representations of "my-body-my-view" space. They will be partly responsible for our ability to derive spatial knowledge from perception, as well as our ability to act in ways that incorporate spatial information of which we are not directly aware (as seen most vividly in deficit cases where patients may be able to point to objects that they aren't aware of seeing or catch a ball without being aware of any visual motion).

These representational structures can be deployed by different motor structures for distinct motor chores, such as moving the eyes, hands, pinnae, legs, or head. The representations are not exactly or merely perceptual, not exactly motor, nor exactly egocentric or allocentric. They combine information from multiple sources in a way that is suited to multiple applications, but cannot neatly be described in terms of a single map or reference frame. Incidentally, this is one of many examples where the representational business of a neuronal pool does not correspond to any folk psychological representation.

Here, as in other places in computational neurobiology, the mathematical workhorse is the parameter space, with mappings across the dimensions of the parameter space yielding coordinate transformations, and volumes or trajectories in parameter space serving as representations. To show why the Pouget and Sejnowski hypothesis is promising for the problem at hand, we must first introduce relevant data concerning parietal cortex, and then explain what is computationally advantageous as well as neurally apropos about their hypothesis.

In the early stages of the visual system (e.g. V1, V2), the location of the visual signal is specified in retinal coordinates. To move the eyes and head to look at a heard or felt object, or to reach with an arm or tongue or foot for a seen object, the brain needs to know where to go in the appropriate coordinate system; retinal coordinates will not suffice. It needs to know, *inter alia*, where the eyeball is with respect to the head, where the head is with respect to the shoulder and trunk. Coordinate transformations are needed to specify where the eyeball should go to foveate, or what, in *joint* coordinates, the arm's position *is* and what it *should be* so that it makes contact with the target (see Figure 4.3). More generally, sensory coordinates have to be transformed into motor coordinates in order to connect with a sensorily specified target.

Integration of somatosensory "body-knowledge", including proprioceptive and vestibular knowledge such as "where this body part is in relation to other body parts", with visual-auditory "where things are in relation to my body" knowledge, allows for general representations of "me in external space". As Damasio has argued, however, the spatial aspects of body-representation are

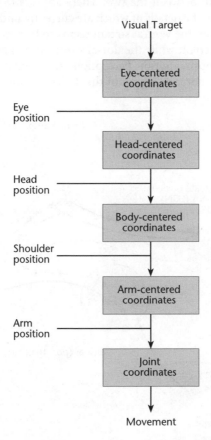

Figure 4.3 Coordinate transformations required to specify an arm movement toward a visual target. From Pouget and Sejnowski (1997c), used by permission of the authors.

only one *part* of the self-representation story, since other aspects involving various dimensions of feeling and homeostasis will figure in what it is to have a "me" representation.

Neurobiological studies of areas 7a and 7b in parietal cortex have provided important clues as to how coordinate transformations are accomplished. What follows is a simplified description of what is known, but it should suffice to convey the main points. As background, note that the visual system in monkeys (and probably primates generally) has a number of important early (pre-cortical) components, including major projections to the superior colliculus and the pulvinar of the thalamus (see Figure 4.4). The pathway we consider here goes from the lateral geniculate nucleus (LGN) of the thalamus to cortical area V1. Thereafter, there is a broad divergence into two broad pathways commonly referred to as the dorsal and ventral streams, with some significant cross-talk between the two. There appears to be divergent functional specialization, the details of which are currently under investigation. To a first approximation, the ventral stream seems to be specialized for categorization, shape, and color, while the dorsal stream shows greater sensitivity to tasks such as stereopsis, motion perception, and spatial location. Area 7 is located in the more anterior region of the dorsal stream. The ventral stream

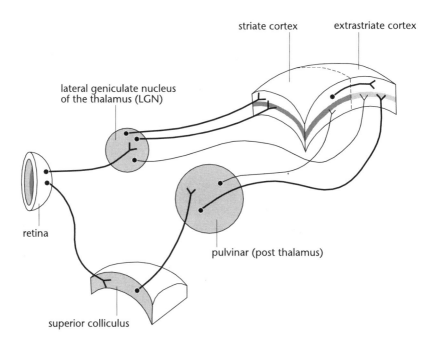

Figure 4.4 Retinocollicular and retinogeniculate visual pathways in mammals. Heavy and thin lines represent major and minor projections. Modified from Butler and Hodos (1996).

appears to play an important role in the perception of visual space as well (Milner and Goodale 1995), and this should be borne in mind despite our focus here on the dorsal stream.

Monkeys with bilateral lesions in area 7 show poor reaching to a target (ataxia), mis-shaping of hands to fit the shape of the target, and slowness of movement. They also show defective eye movements, principally in foveating, and they have other impaired spatial abilities. They are poor at finding the home cage when released, poor at route finding to a food source, and poor in judging spatial relationships among objects (e.g. "the food source is the box located nearer to the can"). Another region of parietal cortex is area 5, which contains cells that fire maximally to a signal when the arm is reaching, and others that fire selectively to the expectation of a stimulus. Because a great deal of research has probed the visual properties of this area, one tends to think of these regions as essentially visual. Recent data reveal, however, that they are much more than that. Response modifiability by many factors including auditory, somatosensory, and vestibular signals, attention, intention, expectation, preparation, and execution, clearly indicates these neurons are more than just sensory. (See Andersen, Essick, and Siegel 1985; Andersen *et al.* 1990; Andersen 1995; Mazzoni and Andersen 1995; Wise *et al.* 1997.)

Area 7 is multimodal, and contains cells responsive individually to either visual, auditory, somatosensory, chemical, vestibular, or proprioceptive signals. Interestingly, auditory cells in this region appear to be mapped in retinotopic coordinates. A few cells are themselves multimodal: they may respond to both visual and auditory signals, or to somatosensory and visual signals, or to chemical and somatosensory signals (Wise *et al.* 1997).

Richard Andersen and his colleagues (1985) discovered that certain ostensibly visual cells show modified responsivity as a function of eyeball position. That is, if a given cell's receptive field was, say, at $+15°, -14°$, then its firing to a signal at that retinal position would be modified depending on where the eyeball was in the head. Crudely, this means that such cells "know" where the external object is *in head-centered space*. The discovery is important because it means that some cells in area 7 are the beneficiaries of a transformation that uses retinal coordinates and eyeball position coordinates to yield information about where external targets are in head coordinates. The representation is thus *somewhat* perceptual, *somewhat* "perceived-thing-relative-to-my-body". Such computational results can then be used by the brain for foveation or reaching. What computation enables coordinate transformations from retinal position and eyeball position to position in head space or body space? How, roughly speaking, can you get *objective* space out of *subjective* spatial relations?

In a classic example of how neural modeling can yield useful ideas for neuroscience, Zipser and Andersen (1988) asked whether an artificial neural net could perform the task of finding the head-centered position of visual objects from their retinal position and the position of the eyeball. Using those two types of position information as input, they trained the net by back-propagation to specify positions in head-centered coordinates. It was a major finding that the two kind of inputs were indeed sufficient for a network to generate a head-centric representation. Exactly what function was computed by the hidden units, however, was not yet clear. Pouget and Sejnowski then investigated the hidden units in the context of a wide range of neurobiological data, including single-cell data. What emerged was a plausible but rather surprising answer to the question of how neurons accomplish the task.

Except for organisms whose sensory and motor systems are *very* simple, the transformation from sensory to motor coordinates is nonlinear. This includes transformations from retinal coordinates to eye position coordinates and joint coordinates and tongue coordinates. As Pouget and Sejnowski note, the retinal receptive fields are Gaussian and the brain does not have access to the retinal position R as a series of numbers—the horizontal and vertical components—but to a set of nonlinear functions of R. Girosi, Jones, and Poggio (1995) had explored the idea that basis functions are an efficient way for the nervous system to approximate nonlinear functions. Gaussians and sigmoids are a subclass of basis functions, as is the product of a Gaussian and a sigmoid.

The Pouget and Sejnowski story goes as follows: eye position units are sigmoidal, and retinal position units are Gaussian. Hidden units (interneurons in area 7) compute the product of the two, which is a basis function (see Figure 4.5). This example involves only two dimensions, but the idea can—and, for neurobiological reasons, *should*—accommodate dimensions for somatosensory, proprioceptive, vestibular, and other signals. Thus, each hidden unit provides a set of basis functions. Computationally this is convenient, because the same basis functions can then be used by different regions to compute the movement appropriate for the appropriate reference frame— reaching or saccadic eye movement or head-turning to an auditory or visual stimulus or whatever. In short, on the Pouget and Sejnowski hypothesis, the area 7 cells studied by Andersen and colleagues are disposed to represent "where perceived objects are in my body space". (See also Andersen *et al.* 1997.)

This result is surprising because it means that rather than actually generating representations for many different reference frames, the brain may use the more down-scale method of generating one set of basis functions and manipulating them as needed. When pushed through relatively simple distinct filters, they provide information suitable to the motor pathway selected (eye muscles, head muscles, trunk muscles, etc.). Depending on the sensorimotor goals, emphasis may be given to cells that are driven primarily by particular sorts of inputs—so a system involved in directed reaching might read off from area 7 a "world" of somatosensory-to-visual coordinations, while a system related to postural adjustments might read off a very different "world" defined by somatosensory, vestibular, and proprioceptive axes (see Figure 4.6).

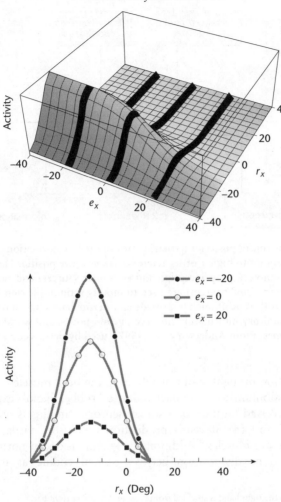

Figure 4.5 Basis function obtained by multiplying a Gaussian of retinal location with a sigmoid of eye position (*top*). When sampled at three different gaze angles (thick lines on top graph), the visual receptive field (*bottom*) shows the same modulation as found in parietal cortex. From Pouget and Sejnowski (1997*c*), used by permission of the authors.

What introspection presents as the "oneness" of spatial perception is undoubtedly illusory to some degree. Various versions of "where perceived objects are in my body space" can dissociate (largely without introspective notice) as a function of precisely which perceptual modalities are involved. The effect has been demonstrated in a variety of experiments. Some examples include: (1) the ventriloquist phenomenon, whereby sound is perceived as coming from an object that merely moves in synchrony with it; (2) changes in

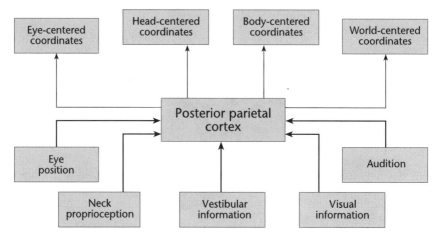

Figure 4.6   The role of posterior parietal cortex in the transformation of retinotopic visual information into higher-order reference frames. Eye position, head position (determined from neck proprioception and vestibular sources) and gaze position (determined from visual sources) are used to modify retinotopic signals. Posterior parietal cortex is thus positioned to provide an intermediate stage in the conversion of visual and auditory information into eye-, head-, body-, and world-centered coordinate frames. From Andersen *et al.* (1997), used by permission of the authors.

visual perception brought about by vibrating the neck muscles (thus stimulating the vestibulum); (3) Ramachandran's "rubber hand" experiment in which the perceived location of a somatosensory stimulus is outside of the actual body;[5] and (4) Stevens's production of illusory visual motion by paralyzing the eye muscles.[6] Additionally, *within* vision, in normal subjects, there can also be dissociation between spatial coding as it is visually

---

[5] In this set-up the right hand is shielded from view, and a rubber right hand is placed in view on a table, about where one might place one's right hand. The subject watches while the experimenter makes many touching, stroking, and tapping motions on the rubber hand. Out of sight, identical motions are made on the subject's real hand. In my (PSC) case, after about three minutes, I ceased to feel the touches as located in my right hand; they felt like they were really located in the rubber hand. The sensation was very robust. The other subject on that day, David Hubel, had a similar experience. Nevertheless it should be noted that not all subjects make the transfer so readily. Although my experience was not the result of a well-controlled experiment, Ramachandran has obtained similar results with subjects under controlled conditions.

[6] See Matin, Stevens, and Picoult 1983. In this experiment, run on himself and his colleagues, Stevens immobilized the eye muscles via a procedure known as a retrobulbar block. When a light is flashed in the visual periphery (e.g. to the right), one intends to move the eyes to the right to look at it. Because the extraocular muscles are paralyzed, however, no eye movement can happen. This mismatch between intent and performance produces the vivid visual *experience* of the whole world making an abrupt jump to the right (as though the eyes really had moved and—since the world still looks the same—the world must have moved along with them). In fact, of course, nothing moved, neither eyes nor world. One simply *intended* the eyes to move. This is a wonderful example of feedback from the motor command trumping visual motion.

experienced and as it is used for grasping (Milner and Goodale 1995).[7] As is well known, such dissociations also occur in patients with parietal lesions (see the discussion of hemineglect, below).

To the extent that the experienced "oneness of space" is not illusory, it is highly dependent on the fact that all of the signals are generated in *one* nervous system, inside *one* body that is *one* spatially linked source of signals. There is no single objective spatial representation (of the sort standardly presupposed by symbolic models of representation), but a distributed, multi-modal representation that fundamentally integrates perception and action, self and world. Where constancies appear across distinct modalities, it becomes possible and even inevitable to understand them as representing an enduring world beyond the body. Kant's "transcendental unity of apper-ception" is turning out to be the fundamental integration of diverse sensory and motor coordinate systems, exquisitely configured *physically* to represent the world.[8]

4. Deficits of Spatial Reasoning

The previous sections examined, from the behavioral and neurological per-spectives, the remarkable powers of spatial representation and reasoning demonstrated by humans and animals alike. To complete our survey of the empirical data, we turn now to an account of some *defects* in spatial reasoning found in humans with parietal lobe damage. It has recently become apparent that nonsymbolic models such as those discussed above can predict the fine structure of these deficits to a remarkable degree.

One intriguing pattern of breakdown in spatial reasoning occurs in *hemi-neglect*, a condition found in patients with unilateral lesions of right parietal cortex. These patients display a marked tendency to ignore the contralesional (i.e. left) side of their body-centered world. They tend to look only to the right, though some can move their eyes to the left if directly asked to do so. Asked to make a drawing or reproduce a figure, they will omit most or all of the left half; asked to "cancel" (cross out) all the lines on a page, they will fail to cancel all the lines on the left half of the page; asked to bisect a horizontal line, they will draw the transecting line off-center to the right.

In one ingenious experiment (Bisiach and Luzzatti 1978), hemineglect patients were asked to imagine they were standing in a well-known plaza in their city, and to describe what they could see from a given vantage point.

---

[7] They use a visual illusion in which a disc visually appears to be larger than it is, but the grasp aperture in *reaching* for the disc is set to the correct size.

[8] On the positive side, Kant's characterization of space as a "form of intuition" might turn out to have a discernible neurophysiological basis, investigation of which could yield new insights into the fundamental structure of human perception. Such considerations also lend weight to Searle's claims about the inadequacy of the functionalist paradigm.

Their descriptions omitted objects on the neglected side *relative to the imagined vantage point*: told to imagine they were standing at the north end of the plaza, they would omit the buildings to the east, and when subsequently instructed to repeat the task from the southern vantage point they would list the eastern buildings and omit from their description all the western buildings that they had listed just moments before. This shows that hemineglect is a deficit of spatial reasoning and/or representation at some fairly basic level, and not just a perceptual failure. Corroborating evidence comes from studies in which a neglect patient perceived rotated non-symmetric figures as normalized (that is, rotated back to their canonical positions) and neglected the *normalized* left half, something which could only happen if the loss of information occurred at a later stage than some fairly high-level aspects of visual processing.

There is also a motor component to hemineglect. Neglect patients show little or no spontaneous use of limbs on the left side of the body, though some will reluctantly move the neglected limbs upon direct request. They will also neglect auditory stimuli from the left, sometimes failing to acknowledge others who are speaking to them from that side. This sort of polymodal, perceptuomotor deficit is what one might expect from a lesion of parietal cortex, since it receives and integrates inputs from multiple modalities and is known to be involved in the coordination of perception with action.

Pouget and Sejnowski (1997a, 1997b, 1997c) used hemineglect as a test of the predictive power of their basis function hypothesis (described in Section 3 above). They created a network model which has the oculomotor input/output structure and response properties that they attribute to area 7a, and then "lesioned" it by removing the units that correspond to the right side of the brain.[9] This left the network with a disproportionately high number of neurons that were most responsive to rightward eye positions and/or right visual field stimuli. They then equipped it with a winner-take-all output selection mechanism, and tested it on stimuli similar to those used with neglect patients.

The network output exhibited striking similarities to the human behavioral results. In the line cancellation task, the network failed to cancel lines on the side opposite the lesion. More importantly, the line between the cancelled and non-cancelled areas was fairly sharp, even though the underlying representation had only a smooth gradient. The network also paralleled human behavior on the bisection task: it was successful before the lesion, and shifted the

---

[9] The units in each hemisphere were organized into maps, with one axis representing sensitivity to horizontal retinal field position (vertical position was not considered) and the other representing sensitivity to eye position. The maps were constructed to have neuronal gradients, such that the right hemisphere had more neurons which were responsive to left retinal field and eye positions, and vice versa for the left hemisphere. Parietal cortex is known to have these sorts of gradients for retinal position; eye position gradients are observed in other areas, but it is not known whether they exist in parietal cortex.

transection point to the right after the lesion. In other experiments, it was shown to suffer object-centered as well as visual-field-centered neglect, and a version that received head position information (instead of eye position) exhibited a curious effect found in human patients, whereby performance on left-field tasks can be improved by turning the head to the right. Both of these phenomena have been problematic for existing theories of hemineglect, and Pouget and Sejnowski are able to explain how they arise naturally from an organization of response functions that can plausibly be attributed to human parietal cortex.[10]

While hemineglect is associated with unilateral damage to parietal cortex, bilateral damage can produce another, more rare condition known as Balint's syndrome. Balint's syndrome patients have little or no ability to locate, count, or track objects, and have poor or nonexistent depth and motion perception. In addition, they exhibit *simultagnosia,* an inability to perform perceptual tasks involving more than one object present in the visual field at the same time.

There is no established explanation of the cluster of deficits seen in Balint's syndrome. One line of theory holds that parietal cortex is centrally involved in the task of disengaging attention from one object so that it can be directed to another, and that patients with parietal damage are thus unable to shift attention between multiple objects (Posner *et al.* 1984; Farah 1990).

More recently, Lynn Robertson, Anne Treisman, and their colleagues have proposed that the central deficit in Balint's has to do with feature binding (Robertson *et al.* 1997). Parietal cortex is the terminus of the dorsal pathway or "*where* system", which extracts information about the spatial position, relations, and movement of visual objects. Based on Treisman's "Feature Integration Theory" (Treisman and Gelade 1980)[11] and on the group's study of a Balint's patient, Robertson and colleagues claim that parietal cortex establishes a spatial map which is required for the proper grouping of perceptual features (as detected by the ventral pathway or "*what* system") into separate, coherent objects.

Robertson and colleagues make a strong experimental case for the claim that Balint's patients suffer from a jumbling-together of the features of the objects in their visual field. They also provide intriguing anecdotal support for

[10] The Pouget and Sejnowski theory is based upon linear combinations of continuous functions. It may be suggested that such a model can (in principle) be approximated to an arbitrary degree of accuracy by a Turing machine. Supposing this to be true, nothing is revealed about which model most accurately captures what the brain is really doing. The behavior of the solar system could also be approximated to an arbitrary degree of accuracy by a Turing machine, but insisting that planetary motion actually involves symbol manipulation according to syntactically specified rules is unrewarding. That Turing equivalence is irrelevant here is further illustrated by the fact that the Pouget and Sejnowski model could also be implemented by analog VLSI, which is about as nonsymbolic as you can get.

[11] A model which, it should be mentioned, was originally developed to explain the behavioral data on feature binding in normals.

this hypothesis. Their patient spontaneously hit upon the technique of looking at individual objects through a tube, which would be a useful way to eliminate the features of other objects (or, on the Posner/Farah accounts, distractors). Commenting on experimental stimuli, he said things like, "It looks like sometimes one letter is going into the other one. I get a double identity. It kind of coincides", and "I can only see two or three [dots] out of the whole bunch."

For our purposes, several things are interesting about the Robertson *et al.* account. First, it is another example of a model based on the known structure of the (healthy) brain which produces good predictions about the behavior of damaged brains. If one of the goals in modeling is to understand the ways in which humans actually perform tasks, then this sort of prediction is clearly an important test (and to date we have been unable to locate a *single* nontrivial prediction of this sort made—let alone made and confirmed—by a symbolic model of spatial representation). Second, though the Robertson *et al.* model does not specify an implementation structure, it is clear that any attempt to implement it as a strongly symbolic system would run directly into the computational explosion that has become famous in AI as the "frame problem" (discussed below), since it calls for the dynamic binding of large numbers of features on a constantly shifting visuospatial array. And finally, it highlights an important fact about human spatial representation, namely that it is fundamentally distributed. Between them, the dorsal and ventral pathways extract and represent all the visuospatial information needed by other areas of the brain, but at no point do they put it together into stable intersubstitutable entities of the sort required for symbolic representation and manipulation.

5.   Spatial Reasoning and the Science of Mind

A central feature of symbolic approaches is that they demand a sharp distinction between the symbolic and non-symbolic powers of the brain. The most common form of the distinction is derived from *functionalism*, the philosophical hypothesis that mental states are defined by their roles in a functional (usually, computational) economy of other such states, independently of their physical instantiations.[12] This approach to the meaning (or "intentionality") of mental states, combined with the computationalist approach to cognition that grew out of the work of Turing and of Newell and Simon, motivated a contrast between levels of description:

What psychology is about is the causal structure of the mind at the intentional level of description. It may very well be, for all I know, that at some neurological level it is a

---

[12] The other common form of the distinction is in terms of modularity. Since all the modules ultimately have to be implemented in wet stuff, though, the problems will be much the same.

system that obeys connectionist postulates. That is of no interest to psychology...the question of what the causal structure of the brain is like at the neurological level might be settled in a connectionist direction and leave entirely open what the causal structure of the brain is like at the intentional level. (Fodor 1995)

There are a number of problems with this approach. For one thing, functionalism has been shown to have severe problems with such basic human functions as sensation, analogical reasoning, and conceptual change, and ultimately it failed in its central goal of providing a functional/computational account of the meanings of words and mental states (which is why it was abandoned by its creator, Hilary Putnam). There remains no respectable argument, in philosophy of mind or in any of the cognitive sciences, for the *a priori* division of the mind/brain into autonomous levels of description. Of course, none of this rules out the *possibility* that our mind/brain just happens to be so divided; let us therefore take this Fodorian claim as an empirical hypothesis, and see how it fares.

Assume for the moment that it *is* worthwhile to draw a line between the intentional, symbolic level of description and the implementational, neural level. What should we say about spatial reasoning? On which side of the line does it belong? In those cases where they have been attempted, symbolic accounts of spatial processing have met with very limited success; even leaving aside questions of their empirical plausibility, such models are haunted by their own personal bugbear, the infamous "frame problem" (See Pylyshyn 1987 for reviews). Simplified, the frame problem says that deductivist representational systems must derive moment-to-moment properties of the represented world from a stored corpus of first-order axioms and a fairly sparse input. Unfortunately for these models, contingency is a fact of life in the real world, and it is essentially unbounded. Brute force prediction of chess moves is a huge, but finite problem. Brute force prediction of large-scale spatial interactions in a highly contingent world is beyond the means of any physical symbolic computer, present or future.[13]

Looking at actual spatial reasoners—ourselves—yields the same conclusion. There doesn't seem to be any scientifically useful level of description at which spatial reasoning is best described as an implementation-independent symbolic process. From single-cell response properties in area 7a, through visual and mental images and conscious control of movement, and all the way

---

[13] Neurally inspired models may handle the problem of contingent response with less difficulty. For instance, it may be convenient to represent even deterministic contingent events as being probabilistic in nature. Probabilistic events present little difficulty for neural nets; even deterministic network models can encode relative probabilities with ease. If there is anything which neural nets are spectacularly good at, it is statistical analysis. Likewise, the problem of relative salience—that is, of predicting which among a vast list of features in a complex environment may determine change—has an elegant solution in the form of the delta rule and other connectionist learning rules. These rules do nothing *but* assign relevance to the different elements of a multidimensional input. Finally, of course, massively parallel architecture solves the problem of determining which elements need to be active at any given moment, by the straightforward expedient of activating them all.

up to mathematical problem-solving and social negotiation, there seems to be a smooth continuity of functional and explanatory dependence. These "levels of analysis" interact with each other, and the character of that interaction has a lot to do with how we see and feel, integrate across modalities, learn and recall, shift attention, and find our way around in space. If there is a line to be drawn, it clearly can't be within or below the domain of spatial reasoning.

Nor does it seem appropriate to draw the line above it, excluding spatial reasoning from the realm of the intentional. When you solve a graphically presented geometry problem by inspection, you don't feel disappointed at having failed to engage in intentional thought; quite the contrary. If the problem was a difficult one, you probably feel pleased with your intellect, and relieved at having solved the problem by cleverness instead of by "brute force" calculation. It is no accident that quarterbacks in (American) football, the players who must quickly evaluate complex spatial patterns composed of intentional agents, must pass an intelligence test before being considered for the major leagues.

It seems, then, that one can find no level of description at which intentional and symbolic thought will be coextensive. Spatial reasoning (along with social reasoning, creativity, and a host of other human functions) gets in the way: at least some intentional thought is spatial, and spatial thought is not best understood in symbolic terms. The functionalist levels-of-analysis hypothesis is simply incompatible with the data.

None of this is to deny that some kinds of thought *are* well described as symbolic. There is at least one form of symbolic activity—language—which all normal humans (and only humans) are astonishingly good at, and many others—algebra, formal logic, chess—which can be learned with greater or lesser degrees of difficulty. These activities display, in an approximate and finite form, many of the properties of genuine symbolic systems.[14] At present there is very little understanding of the way in which the brain conducts these operations, and we regard this as a deep and pressing issue in cognitive science. Symbol manipulation is a strange and wonderful power; understanding how we do it would be a transformative advance in the science of mind.[15]

What we do deny is that the way to understand symbolic cognition—or for that matter, *any* form of cognition—is to divorce it from other forms of thought and from the body in which it occurs. In the face of the work discussed earlier, and a host of other studies conducted by Ramachandran, the Damasios, and others, it seems incontestable that studying the brain can provide deep insight into the nature of the mind. To be sure, this doesn't mean that every theory in the cognitive sciences has to make mention of the brain; but it does mean that actively *denying* the relevance of the body and

[14] Which should come as no surprise, since the nonhuman symbolic systems with which we are now so familiar—from the hypothetical Turing machine through the desktop computer—were created in the image of these human competencies.

[15] For one recent attempt, see Deacon (1997).

brain—to *any* cognitive function, however "high" or abstract—can only hinder the progress of understanding.

Why, then, do so many symbolic computationalists insist on just such a denial? Why do they feel compelled to claim sole dominion over the task of explaining the human mind (or at least, the good parts), instead of acknowledging that the many levels of description and analysis should be richly interpenetrated and mutually informative?

The answer lies in the motivation which drove the development of mechanical symbol systems. As Clark (1993) puts it, a physical symbol system—however instantiated—must be "semantically well-behaved". It must be so constituted that some aspects of its physical states can be mapped systematically onto semantic states, and that the transformations in its physical states map onto legal moves in semantic space. For example, in the case of a calculator, this means that each of its (normally reachable, electronic) states must have some mathematical meaning, and that the effect of pressing a button should be to bring about a new state that is mathematically related to the previous state.

What makes semantically well-behaved systems so useful is that they're *truth-preserving*. If you start with a state that represents some true fact, then proper manipulations of the system will always result in further true facts, never in falsehoods. In short, these systems are reliable tools for doing deductive logic and mathematics. It is clear why one might hope that the human brain would turn out to be such a system: it would provide a natural explanation of our formal abilities; it would give us a reason to believe that the deep structure of our thought will be amenable to analysis using the tools and vocabulary that are available to us now; it would even imply heartening things about the educability and rationality of human beings. But there is a catch: to guarantee truth-preservation,[16] a system has to be *purely* symbolic. If it overlaps or interacts (beyond the level of input/output relations) with some nonsymbolic system, then its state transitions will not be governed solely by their syntactic relations, and thus their semantic validity may be compromised.

The problem is, when you look at the actual human mind/brain, truth-preservation *in this formal sense* does not seem to be much of a priority. Both in neuroscience and in psychology, human beings turn out to be chock full of useful fictions. Human perception and reasoning are shot through with domain-specific heuristics and short-cuts, with confabulations, with undischarged assumptions (and if any of our readers find this characterization implausible, we recommend that they try their hand at teaching an introductory course in formal logic). Wherever we look in the brain, we find systems whose structure is very ill-suited to guarding against the

---

[16] And other beloved properties such as compositionality, the ability to "nest" statements to arbitrary depth by substituting complex expressions for simple symbols.

possibility of falsehood but very well-suited to swift and flexible perception and action.

What we have tried to show is that there is a poor fit between the actual structure of the human mind/brain and the central motivating principles of the symbolic approach. This does not mean that symbolic characterizations of thought are useless; they still constitute the only general, systematic, inter-subjective way of compactly describing high-level thought. What it does mean is that we should not accept any claims about the ubiquity or autonomy of symbolic processes. Thinking in terms of the brain's more pragmatic goals provides a better framework for asking questions, building models, and understanding the mental and physical structure of thought.

5 Toward a Cognitive Neurobiology of the Moral Virtues

Paul M. Churchland

1. Introduction

These are the early days of what I hope will be a long and fruitful intellectual tradition, a tradition fueled by the systematic interaction and mutual information of cognitive neurobiology on the one hand and moral theory on the other. More specifically, it is the traditional sub-area we call *meta*ethics, including moral epistemology and moral psychology, that will be most dramatically informed by the unfolding developments in cognitive neurobiology. And it is metaethics again that will exert a reciprocal influence on future neurobiological research—more specifically, into the nature of moral perception, the nature of practical and social reasoning, and the development and occasional corruption of moral character.

This last point about reciprocity highlights a further point. What we are contemplating here is no imperialistic takeover of the moral by the neural. Rather, we should anticipate a mutual flowering of *both* our high-level conceptions in the domain of moral knowledge *and* our lower-level conceptions in the domain of normal and pathological neurology. For each level has much to teach the other, as this essay will try to show.

Nor need we resist this interaction of distinct traditions on grounds that it threatens to deduce normative conclusions from purely factual premises, for it threatens no such thing. To see this clearly, consider the following parallel. Cognitive neurobiology is also in the process of throwing major illumination on the philosophy of *science*—by way of revealing the several forms of neural representation that underlie scientific cognition, and the several forms of neural activity that underlie learning and conceptual change (see e.g. P. M. Churchland 1989*b*, 1989*c*, 1989*d*). And yet, substantive science itself will still have to be done by scientists, according to the various methods by which we make scientific progress. An adequate theory of the brain, plainly, would not constitute a theory of Stellar Evolution or a theory of the underlying structure of the Periodic Table. It would constitute, at most, only a theory of how we generate, embody, and manipulate such worthy cognitive achievements.

Equally, and for the same reasons, substantive moral and political theory will still have to be done by moral and political thinkers, according to the various methods by which we make moral and political progress. An adequate

This chapter was first published in *Topoi*, 17 (1998), 77–82. Reprinted with kind permission of *Topoi*.

theory of the brain, plainly, will not constitute a theory of Distributive Justice or a body of Criminal Law. It would constitute, at most, only a theory of how we generate, embody, and manipulate such worthy cognitive achievements.

These reassurances might seem to rob the contemplated program of its interest, at least to moral philosophers, but we shall quickly see that this is not the case. For we are about to contemplate a systematic and unified account, sketched in neural-network terms, of the following phenomena: moral knowledge, moral learning, moral perception, moral ambiguity, moral conflict, moral argument, moral virtues, moral character, moral pathology, moral correction, moral diversity, moral progress, moral realism, and moral unification. This collective sketch will serve at least to outline the program, and even at this early stage it will provide a platform from which to address the credentials of one prominent strand in pre-neural metaethics, the program of so-called "Virtue Ethics", as embodied in both an ancient writer (Aristotle), and three modern writers (Johnson, Flanagan, and MacIntyre).

2. The Reconstruction of Moral Cognitive Phenomena in Cognitive Neurobiological Terms

This essay builds on work now a decade or so in place, work concerning the capacity of recent neural-network models (of micro-level brain activity) to reconstruct, in an explanatory way, the salient features of molar-level cognitive activity. That research began in the mid-1980s by addressing the problems of perceptual recognition, motor-behavior generation, and other basic phenomena involving the gradual learning of sundry cognitive *skills* by artificial "neural" networks, as modeled within large digital computers (Gorman and Sejnowski 1988; Lehky and Sejnowski 1988; Rosenberg and Sejnowski 1990; Lockery, Fang, and Sejnowski 1991; Cottrell 1991; Elman 1992). From there, it has moved both downward in its focus, to try to address in more faithful detail the empirical structure of biological brains (P. S. Churchland and Sejnowski 1992), and upward in its focus, to address the structure and dynamics of such higher-level cognitive phenomena as are displayed, for example, in the human pursuit of the various theoretical sciences (P. M. Churchland 1989*a*).

For philosophers, perhaps the quickest and easiest introduction to these general ideas is the highly pictorial account in P. M. Churchland (1995), to which I direct the unprepared reader. My aim here is not to recapitulate that groundwork, but to build on it. Even so, that background account will no doubt slowly emerge, from the many examples to follow, even for the reader new to these ideas, so I shall simply proceed and hope for the best.

The model here being followed is my earlier attempt to reconstruct the epistemology of the *natural* sciences in neural-network terms (P. M. Churchland 1989*a*). My own philosophical interests have always been centered around issues in epistemology and the philosophy of science, and so it was natural, in the mid-1980s, that I should first apply the emerging framework of cognitive neurobiology to the issues with which I was most familiar. But it soon became obvious to me that the emerging framework had an unexpected generality, and that its explanatory power, if genuine at all, would illuminate a much broader range of cognitive phenomena than had so far been addressed. I therefore proposed to extend its application into other cognitive areas such as mathematical knowledge, musical knowledge, and moral knowledge. (Some first forays appear in chapters 6 and 10 of P. M. Churchland 1995.) These further domains of cognitive activity provide, if nothing else, a series of stiff *tests* for the assumptions and explanatory ambitions of neural-network theory. Accordingly, the present essay presumes to draw out the central theoretical claims, within the domain of metaethics, to which a neural-network model of cognition commits us. It is for the reader, and especially for professional moral philosophers themselves, to judge whether the overall portrait that results is both explanatorily instructive and faithful to moral reality.

Moral Knowledge

Broadly speaking, to teach or train any neural network to embody a specific cognitive capacity is gradually to impose a specific *function* onto its input–output behavior. The network thus acquires the ability to respond, in various but systematic ways, to a wide variety of potential sensory inputs. In a simple, three-layer feedforward network with fixed synaptic connections (Figure 5.1*a*), the output behavior at the third layer of neurons is completely determined by the activity at the sensory input layer. In a (biologically more realistic) *recurrent* network (Figure 5.1*b*), the output behavior is jointly determined by sensory input *and* the prior dynamical state of the entire network. The purely feedforward case yields a cognitive capacity that is sensitive to spatial patterns but blind to temporal patterns or to temporal context; the recurrent network yields a capacity that is sensitive to, and responsive to, the changing cognitive contexts in which its sensory inputs are variously received. In both cases, the acquired cognitive capacity actually resides in the specific configuration of the many synaptic *connections* between the neuronal layers, and learning that cognitive capacity is a matter of slowly adjusting the size or "weight" of each connection so that, collectively, they come to embody the input–output function desired. On this, more in a moment.

Evidently, a trained network has acquired a specific skill. That is, it has learned how to respond, with appropriate patterns of neural activity across its output layer, to various inputs at its sensory layer. Accordingly, and as with all other kinds of knowledge, my first characterization of moral knowledge portrays it as a *set of skills*. To begin with, a morally knowledgable adult has

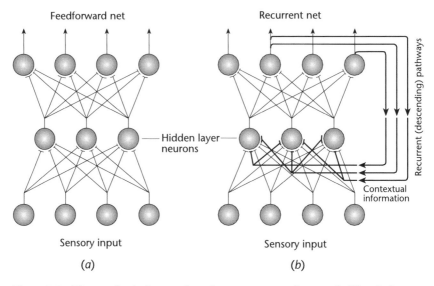

Figure 5.1   The two basic forms of an elementary neural network. The circles represent the cell bodies of neurons; the connecting lines represent axons and their divergent end-branches; the tiny arcs abutting the receiving neurons represent synaptic connections of various sizes. Information flows upwards, from the initial population of sensory neurons, through successive populations of neurons. It is transformed at each stage as it passes through the intervening "filter" of synaptic connections.

clearly acquired a sophisticated family of *perceptual* or *recognitional* skills, which skills allow him a running comprehension of his own social and moral circumstances, and the social and moral circumstances of the others in his community. Equally clearly, a morally knowledgable adult has acquired a complex set of *behavioral* and *manipulational* skills, which skills make possible his successful social and moral interaction with the others in his community.

According to the model of cognition here being explored, the skills at issue are embodied in a vast configuration of appropriately weighted synaptic connections. To be sure, it is not intuitively obvious how a thousand, or a billion, or a trillion such connections can constitute a specific cognitive skill, but we begin to get an intuitive grasp of how they can do so when we turn our attention to the collective behavior of the neurons at the layer to which those carefully configured connections happen to attach.

Consider, for example, the second layer of the feedforward network in Figure 5.1*a*. That neuronal population, like any other discrete neuronal population, represents the various states of the world with a corresponding variety of *activation patterns* across that entire population. That is to say, just as a pattern of brightness levels across the 200,000 pixels of your familiar TV screen can represent a certain two-dimensional scene, so can the pattern of activation levels across a neuronal population represent specific aspects of the external world, although the "semantics" of that representational relation will only rarely be so obviously "pictorial". If the neuronal representation is auditory, for example, or olfactory, or gustatory, then obviously the representation will be something other than a 2–D "picture".

What is important for our purposes is that the abstract *space* of *possible* representational patterns, across a given neuronal population, slowly acquires, in the course of training the synapses, a specific structure—a structure that assigns a family of dramatically preferential abstract *locations*, within that space, in response to a preferred family of distinct stimuli at the network's sensory layer. This is how the mature network manages to categorize all possible inputs, either as rough instances of one-or-other of its learned family of prototypical *categories*, or, failing that, as instances of unintelligible noise. Before training, *all* inputs produce noise at the second layer. After training, however, that second layer has become preferentially sensitized to a comparatively tiny subset of the vast range of possible input patterns (most of which are never encountered). Those "hot-button" input patterns, whenever they occur, are subsequently assimilated to the second layer's acquired set of *prototypical categories*.

Consider an artificial network (Figure 5.2a) trained to discriminate human faces from nonfaces, male faces from female faces, and a handful of named individuals as presented in a variety of distinct photographs. As a result of that training, the abstract space of *possible* activation patterns across its second neuronal layer has become *partitioned* (Figure 5.2b), first into a pair of complementary subvolumes for neuronal activation patterns that represent sundry faces and nonfaces respectively. The former subvolume has become further partitioned into two mutually exclusive subvolumes for male faces and female spaces respectively. And within each of these two subvolumes there are proprietary "hot-spots" for each of the named individuals that the network learned to recognize during training.

Following this simple model, the suggestion here advanced is that our capacity for *moral* discrimination also resides in an intricately configured matrix of synaptic connections, which connections also partition an abstract conceptual space, at some proprietary neuronal layer of the human brain, into a hierarchical set of categories, categories such as "morally significant" vs. "morally nonsignificant" actions; and within the former category, "morally bad" vs. "morally praiseworthy" actions; and within the former subcategory, sundry specific categories such as "lying", "cheating", "betraying", "stealing", "tormenting", "murdering", and so forth. (see Figure 5.3).

That abstract space of possible neuronal-activation patterns is a simple model for our own conceptual space for moral representation, and it displays an intricate structure of similarity and dissimilarity relations; relations that cluster similar vices close together and similar virtues close together; relations

Paul M. Churchland

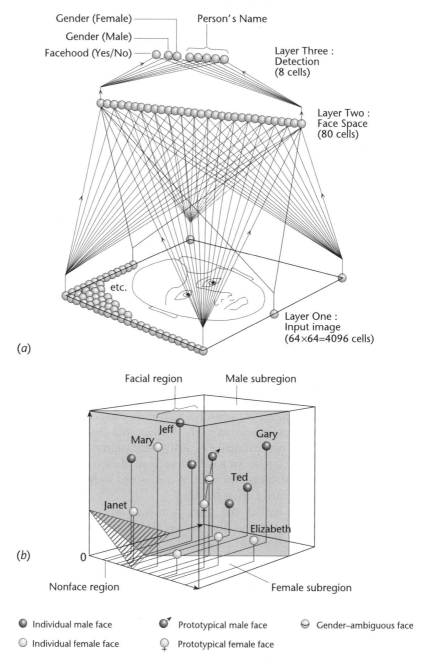

(a)

(b)

⬤ Individual male face   ◉ Prototypical male face   ◒ Gender–ambiguous face

◯ Individual female face   ⊕ Prototypical female face

Figure 5.2 (a) A feedforward network with the capacity to discriminate faces from nonfaces, to discriminate female faces from male faces, and to identify, across a diversity of distinct photographs, each of the eleven individual faces on which it was

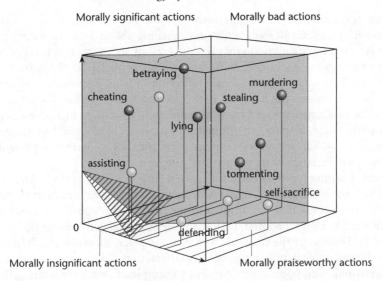

Figure 5.3 A (wholly conjectural) activation space for a possible neural network trained to discriminate morally significant from morally insignificant actions; to discriminate morally bad from morally praiseworthy actions; and to identify each of the salient types of social behavior on which it was initially trained.

that separate highly dissimilar action categories into spatially distant sectors of the space. This high-dimensional similarity space (of course, Figure 5.3 ignores all but three of its many neuronal axes) displays a structured family of categorial "hot spots" or "prototype positions", to which actual sensory inputs are assimilated with varying degrees of closeness.

An abstract space of *motor*-neuron activation patterns will serve a parallel function for the generation of actual social behavior, a neuronal layer that presumably enjoys close functional connections with the sensory neurons just described. All told, these structured spaces constitute our acquired knowledge of *the structure of social space*, and *how to navigate it effectively*.

Moral Learning

Moral learning consists in the gradual generation of these internal perceptual and behavioral prototypes, a process that requires repeated exposure to, or practice of, various *examples* of the perceptual or motor categories at issue. In artificial neural networks, such learning consists in the repeated adjustment of the weights of their myriad synaptic connections, adjustments that are guided

---

trained. (b) A cartoon representation of the space of possible activation patterns across the middle layer of neurons after training on many examples of input images. That space has become partitioned into a hierarchy of perceptual categories.

by the naive network's initial performance *failures*, as measured by a distinct "teacher" program. In living creatures, learning also consists in the repeated adjustment of one's myriad synaptic connections, a process that is also driven by one's ongoing experience with failure. Our artificial "learning technologies" are currently a poor and pale reflection of what goes on in real brains, but in both cases—the artificial networks and living brains—those gradual synaptic readjustments lead to an appropriately structured high-dimensional similarity space, a space partitioned into a hierarchical family of categorial subspaces, which subspaces contain a central hot spot that represents a *prototypical* instance of its proprietary category.

Such learning typically takes time, often large amounts of time. And as the network models have also illustrated, such learning often needs to be structured, in the sense that the simplest of the relevant perceptual and behavioral skills need to be learned first, with the more sophisticated skills being learned later, and only after the elementary ones are in place. Moreover, such learning can display some familiar pathologies, those that derive from a narrow or otherwise skewed population of training examples. In such cases, the categorial framework duly acquired by the network fails to represent the full range and true structure of the social/moral domain it needs to represent, and performance failures are the inevitable result.

These remarks barely introduce the topic of moral learning, but we need to move on. The topic will be readdressed below, when we discuss moral progress.

Moral Perception

This most fundamental of our moral skills consists in the *activation*, at some appropriate layer of neurons at least half a dozen synaptic connections away from the sensory periphery, of a specific pattern of neuronal excitation-levels that is sufficiently close to some already learned moral *prototype* pattern. That nth-layer activation pattern is jointly caused by the current activation pattern across one or more of the brain's sensory or input layers, and by the series of carefully trained synaptic connections that intervene. Moral perception is thus of a piece with perception generally, and its profile displays features long familiar to perceptual psychologists.

For example, one's spontaneous judgments about the social and moral configuration of one's current environment are strongly sensitive to contextual features, to collateral information, and to one's current interests and focus of attention. Moral perception is thus subject to "priming effects" and "masking effects". As well, moral perception displays the familiar tendency of cognitive creatures to "jump to conclusions" in their perceptual interpretations of partial or degraded perceptual inputs. Like artificial networks, we humans have a strong tendency automatically to assimilate our current perceptual circumstances to the nearest of the available moral prototypes that our prior training has created in us.

Moral Ambiguity

A situation is morally ambiguous when it is problematic by reason of its tendency to activate *more than one* moral prototype, prototypes that invite two incompatible or mutually exclusive subsequent courses of action. In fact, and to some degree, ambiguity is a chronic feature of our moral experience, partly because the social world is indefinitely complex and various, and partly because the interests and collateral information each of us brings to the business of interpreting the social world differ from person to person and from occasion to occasion. The recurrent or descending pathways within the brain (illustrated, in stick-figure form, in Figure 5.1*b*) provide a continuing stream of such background information (or misinformation) to the ongoing process of perceptual interpretation and prototype activation. Different "perceptual takes", on one and the same situation, are thus inevitable. Which leads us to our next topic.

Moral Conflict

The activation of distinct moral prototypes can happen in two or more distinct individuals confronting the same situation, and even in a single individual, as when some contextual feature is alternately magnified or minimized and one's overall perceptual take flips back and forth between two distinct activation patterns in the neighborhood of two distinct prototypes. In such a case, the single individual is morally conflicted ("Shall I *protect* a friend's feelings by keeping silent on someone's trivial but hurtful slur, or shall I be forthright and *truthful* in my disclosures to a friend?").

*Inter*personal conflicts about the moral status of some circumstance reflect the same sorts of divergent interpretations, driven this time by interpersonal divergences in the respective collateral information, attentional focus, hopes and fears, and other contextual elements that each perceiver brings to the ambiguous situation. Occasional moral conflicts are thus possible, indeed, they are inevitable, even between individuals who had identical moral training and who share identical moral categories.

There is, finally, the extreme case where moral judgment diverges because the two conflicting individuals have fundamentally different moral conceptual frameworks, reflecting major differences in the acquired structure of their respective activation spaces. Here, even communication becomes difficult, and so does the process by which moral conflicts are typically resolved.

Moral Argument

On the picture here being explored, the standard conception of moral argument as the formal deduction of moral conclusions from shared moral premises starts to look procrustean in the extreme. Instead, the administration and resolution of moral conflicts emerges as a much more dialectical process whereby the individuals in conflict take turns highlighting or making salient certain aspects of the situation at issue, and take turns urging various similarities between the situation at issue and various shared prototypes, in hopes of producing, within their adversary, an activation pattern that is closer to the prototype being defended ("It's a mindless clutch of cells, for heaven's sake! The woman is not obliged to preserve or defend it.") and/or farther from the prototype being attacked ("No, it's a miniature person! Yes, she is obliged."). It is a matter of nudging your interlocutor's current neuronal activation-point *out* of the attractor-category that has captured it, and *into* a distinct attractor-category. It is a matter of trying to change the probability, or the robustness, or the proximity to a shared neural prototype-pattern, of your opponent's neural behavior.

In the less tractable case where the opponents fail to share a common family of moral prototypes, moral argument must take a different form. I postpone discussion of this deeper form of conflict until the section on moral progress, below.

Moral Virtues

These are the various skills of social *perception*, social *reflection*, *imagination*, and *reasoning*, and social *navigation* and *manipulation* that normal social learning produces. In childhood, one must come to appreciate the high-dimensional background structure of social space—its offices, its practices, its prohibitions, its commerce—and one must learn to recognize its local configuration swiftly and reliably. One must also learn to recognize one's own current position within it, and the often quite different positions of others. One must learn to anticipate the normal unfolding of this ongoing commerce, to recognize and help repair its occasional pathologies, and to navigate its fluid structure while avoiding social disasters, both large and small. All of this requires skill in divining the social perceptions and personal interests of others, and skill in manipulating and negotiating our collective behavior.

Being skills, such virtues are inevitably acquired rather slowly, as will be familiar to anyone who has raised children. Nor need their continued development ever cease, at least in individuals with the continued opportunities and the intelligence necessary to refine them. The acquired structures within one's neuronal activation spaces—both perceptual and motor—can continue to be sculpted by ongoing experience and can thus pursue an ever deeper insight into, and an effectively controlling grasp of, one's enclosing social reality. Being skills, they are also differently acquired by distinct individuals, and they are differentially acquired within a single individual. Each brain is slightly different from every other in its initial physical structure, and each brain's learning history is unique in its myriad details. No two of us are identical in the profile of skills we acquire, which raises our next topic.

Moral Character

A person's unique moral character is just the individual profile of his perceptual, reflective, and behavioral skills in the social domain. From what has already been said, it will be evident that moral character is distinguished more by its rich diversity across individuals than by its monotony. Given the difficulty in clearly specifying any canonical profile as being uniquely ideal, this is probably a good thing. Beyond the unending complexity of social space, the existence of a diversity of moral characters simply reflects a healthy tendency to explore that space and to explore the most effective styles of navigating it. By this, I do not mean to give comfort to moral nihilists. That would be to deny the reality of social learning. What I am underwriting here is the idea that long-term moral learning across the human race is positively served by tolerating a Gaussian distribution of well-informed "experiments" rather than by insisting on a narrow and impossible orthodoxy.

This view of the assembled moral virtues as a slowly acquired network of skills also contains an implicit critique of a popular piece of romantic nonsense, namely, the idea of the "sudden convert" to morality, as typified by the "tearful face of the repentant sinner" and the post-baptismal "born-again" charismatic Christian. Moral character is not something—is *not remotely* something—that can be acquired in a day by an Act of Will or by a single Major Insight.

The idea that it can be so acquired is a falsifying reflection of one or other of two familiar conceptions of moral character, herewith discredited. The first identifies moral character with the acceptance of a canonical set of behavior-guiding rules. The second identifies moral character with a canonical set of desires, such as the desire to maximize the general happiness, and so on. Perhaps one can embrace a set of rules in one cathartic act, and perhaps one can permanently privilege some set of desires by a major act of will. But neither act can result in what is truly needed, namely, an intricate set of finely honed perceptual, reflective, and sociomotor skills. These take several decades to acquire. Epiphanies of moral commitment can mark, at most, the initiation of such a process. Initiations are welcome, of course, but we do not give children a high-school diploma for showing up for school on the first day of the first grade. For the same reasons, "born-again" moral characters should probably wait a similar period of time before celebrating their moral achievement or pressing their moral authority.

Moral Pathology

This is a large topic, since, if there are many different ways to succeed in being a morally mature creature, there are even more ways in which one might fail. But as a first pass, moral pathology consists in the partial absence, or subsequent corruption, of the normal constellation of perceptual, reflective, and behavioral skills under discussion. In terms of the cognitive theory that underlies the present approach, it consists in the failure to achieve, or subsequently to activate normally, a suitable hierarchy of moral prototypes within one's neuronal activation space. And at the lowest level, this consists in a failure, either early or late, to achieve and maintain the proper configuration of the brain's 10^{14} synaptic weights, the configuration that sustains the desired hierarchy of prototypes and makes possible their appropriate activation.

The terms "normally", "suitable", "proper", and "appropriate" all appear in this quick characterization, and they will all owe their sense to an inextricable mix of *functional* understanding within cognitive neurobiology and genuine *moral* understanding as brought to bear by common sense and the civil and criminal law. The point here urged is that we can come to understand how displays of moral incompetence, both major and minor, are often the reflection of specific functional failures, both large and small, within the brain. This is not a speculative remark. Thanks to the increasing availability of brain-scanning technologies such as Positron Emission Tomography (PET) and Magnetic Resonance Imaging (MRI), neurologists are becoming familiar with a variety of highly specific forms of brain damage that display themselves in signature forms of cognitive failure in moral perception, moral reasoning, and social behavior (Damasio, Tranel, and Damasio 1991; Damasio 1994; Bechara *et al.* 1994; Adolphs *et al.* 1996; Damasio 1996).

Two quick examples will illustrate the point. The neurologists Antonio and Hanna Damasio have a patient, known in the literature as "Boswell", who is independently famous for his inability to lay down any new long-term memories because of bilateral lesions to his medial temporal lobe, including his hippocampus. Since his illness, his "remembered past" is a moving window that reaches back no more than forty seconds. More importantly, for our purposes, it later emerged that he also displays a curious inability to "see evil" in pictures of various emotionally charged and potentially violent scenes. In particular, he is unable to pick up on the various negative emotions as expressed in people's *faces*, and he will blithely confabulate innocent explanations of the socially and morally problematic scenes shown him. There is nothing wrong with Boswell's eyes or visual system, however. His cognitive deficit lies roughly a dozen synaptic steps and a dozen neuronal layers behind his retinas.

As the MRI scans revealed, Boswell's herpes-simplex encephalitis had also damaged the lower half of both of his temporal lobes, which includes the area called "IT" (infero-temporal) known for its critical role in discriminating individual human faces and in coding facial expressions. He can no longer recognize the identity of faces well-known to him before the illness (movie stars and presidents, for example), and his moral perception has been selectively impaired in the manner described.

A second patient, EVR, had a normal life as a respected accountant, devoted father, and faithful husband. In his mid-40s, a ventromedial frontal brain tumor was successfully removed, and subsequent tests revealed no change in his original IQ of 140. But within six months he had lost his job for rampant irresponsibility, made a series of damaging financial decisions, was divorced by his frustrated wife, briefly married and then was left by a prostitute, and had generally become incapable of the normal prudence that guides complex planning and intricate social interactions. Subsequent MRI scans confirmed that the surgical removal of the original tumor had lesioned the ventromedial frontal cortex (the seat of complex planning) and its connections to the amygdala (a primitive limbic area that apparently embodies fear and anxiety).

The functional consequence of this break in intra-brain communication was to *isolate* EVR's practical reasonings from the "visceral" somatic and emotional reactions that normally accompany the rational evaluation of practical alternatives. In normals, those "somatic markers" (as the Damasios have dubbed them) constitute an important dimension of socially relevant information and a key factor in inhibiting one's decisions. In EVR, they have been cut out of the loop, resulting in the sorts of behavior described above.

These two failures, of moral perception and moral behavior respectively, resulted from sudden illness and consequent damage to specific brain areas, which is what brought them to the attention of the medical profession and led to their detailed examination. But these and many other neural deficits can also appear slowly, as a result of developmental misadventures and other chronic predations—childhood infections, low-level toxins, abnormal metabolism, abnormal brain chemistry, abnormal nutrition, maternal drug use during pregnancy, and so forth. There is no suggestion, let me emphasize, that all failures of moral character can be put down to structural deficits in the brain. A proper moral education—that is, a long stretch of intricate socialization—remains a necessary condition on acquiring a well-formed moral character, and no doubt the great majority of failures, especially the minor ones, can be put down entirely to sundry inadequacies in that process.

Even so, the educational process is thoroughly entwined with the developmental process and deeply dependent on the existence of normal brain structures to embody the desired matrix of skills. At least some failures of moral character, therefore, and especially the most serious failures, are likely to involve some confounding disability or marginality at the level of brain structure and/or physiological activity. If we wish to be able wisely to address such major failures of moral character, in the law and within the correctional system, we would therefore do well to understand the many dimensions of neural failure that can collectively give rise to them. We can't fix what we don't understand.

Moral Correction

Consider first the structurally and physiologically *normal* brain whose formative social environment fails to provide a normal moral education. The child's experience may lack the daily examples of normal moral behavior in others, it may lack opportunities to participate in normal social practices, it may fail to see others deal successfully and routinely with their inevitable social conflicts, and it may lack the normal background of elder sibling and parental correction of its perceptions and its behavior. For the problematic young adult that results, moral correction will obviously consist in the attempt somehow to make up a missed or substandard education.

That can be very difficult. The cognitive plasticity and eagerness to imitate found in children is much reduced in a young adult. And a young adult cannot easily find the kind of tolerant community of innocent peers and wise elders that most children are fortunate to grow up in. Thus, not one but two important windows of opportunity have been missed.

The problem is compounded by the fact that children in the impoverished social environments described do not simply fail to learn. Rather, they may learn quite well, but *what* they learn is a thoroughly twisted set of social and moral prototypes and an accompanying family of skills which—while crudely functional within the impoverished environment that shaped them, perhaps—are positively *dys*functional within the more coherent structure of society at large. This means that the young adult has some substantial *un*learning to do. Given the massive cognitive "inertia" characteristic even of normal humans, this makes the corrective slope even steeper, especially when young adult offenders are incarcerated in a closely-knit prison community of similarly twisted social agents.

This essay was not supposed to urge any substantive social or moral policies, but those who do trade in such matters may find relevant the following purely factual issues. America's budget for state and federal prisons is said to be somewhat larger than its budget for *all* of higher education, for its élite research universities, its massive state universities, its myriad liberal arts colleges, and all of its technical colleges and two-year junior colleges combined. It is at least conceivable that our enormous penal-system budget might be more wisely spent on prophylactic policies aimed at raising the quality of the social environment of disadvantaged children, rather than on policies that struggle, against much greater odds, to repair the damage once it is done.

A convulsive shift, of course, is not an option. Whatever else our prisons do or do not do, they keep at least some of the dangerously incompetent social agents and the outright predators off our streets and out of our social commerce. But the plasticity of the young over the old poses a constant invitation to shift our corrective resources childwards, as due prudence dictates. This policy suggestion hopes to reduce the absolute input to our correctional institutions. An equally important issue is how, in advance of such "utopian" advances, to increase the rate at which they are emptied, to which topic I now turn.

A final point, in this regard, about normals. The cognitive plasticity of the young—that is, their unparalleled capacity for learning—is owed to neuro-chemical and physiological factors that fade with age. (The local production and diffusion of nitric oxide within the brain is one theory of how some synaptic connections are made selectively subject to modification, and there are others.) Suppose that we were to learn how to *recreate* in young adults, temporarily and by neuropharmacological means, that perfectly normal regime of neural plasticity and learning aptitude found in children. In con-junction with some more effective programs of resocialization than we cur-rently command (without them, the pharmacology will be a waste of time), this might re-launch the "disadvantaged normals" into something much closer to a normal social trajectory and out of prison for good.

There remain, alas, the genuine abnormals, for whom moral correction is first a matter of trying to repair or compensate for some structural or physiological defect(s) in brain function. Even if these people are hopeless, it will serve social policy to identify them reliably, if only to keep them permanently incarcerated or otherwise out of the social mainstream. But some, at least, will not be hopeless. Where the deficit is biochemical in nature—giving rise to chronically inappropriate emotional profiles, for ex-ample—neuropharmacological intervention, in the now-familiar form of chronic subdural implants, perhaps, will return some victims to something like a normal neural economy and a normal emotional profile. That will be benefit enough, but they will then also be candidates for the resocialization techniques imagined earlier for disadvantaged normals.

This discussion presumes far more neurological understanding than we currently possess, and is plain speculative as a result. But it does serve to illustrate some directions in which we might well wish to move, once our early understanding here has matured. In any case, I shall close this discussion by reemphasizing the universal importance of gradual socialization by long interaction with a moral order already in place. We will never create moral character by medical intervention alone. There are too many trillions of synaptic connections to be appropriately weighted and only long experience can hope to do that superlatively intricate job. The whole point of exploring the technologies mentioned above will be to maximize everyone's chances of engaging in and profiting from that traditional and irreplaceable process.

Moral Diversity

I here refer not to the high-dimensional bell-curve diversity of moral characters within a given culture at a given time, but to the nonidentity, across two cultures separated in space and or in time, of the overall *system* of moral prototypes and prized skills common to most normal members of each. Such major differences in moral consciousness typically reflect differences in substantive economic circumstances between the two cultures, in the peculiar threats to social order with which they have to deal, in the technologies they command, the metaphysical beliefs they happen to hold, and other accidents of history. Such diversity, when discovered, is often seen as grounds for a blanket skepticism about the objectivity or reality of moral knowledge. That was certainly its effect on me in my later childhood, a reaction reinforced by the astonishingly low level of moral argument I would regularly hear from my more religious schoolchums, and even from the local pulpits. But that is no longer my reaction, for throughout history there have been comparable differences, between distinct cultures, where *scientific* knowledge was concerned, and comparable block-headedness in purely "factual" reasoning (think of "New Age medicine", for example, or "UFOlogy"). But this very real diversity and equally lamentable sloppiness does not underwrite a blanket skepticism about the possibility of scientific knowledge. It merely shows that it is not easy to come by, and that its achievement requires a long-term process of careful and honest evaluation of a wide variety of complex experiments over a substantial range of human experience. Which points to our next topic.

Moral Progress

If it exists—there is some dispute about this—moral progress consists in the slow change and development, over historical periods, of the moral prototypes we teach our children and forcibly impose on derelict adults, a developmental process that is gradually instructed by our collective *experience* of a collective life lived under those perception-shaping and behavior-guiding prototypes.

From the neurocomputational perspective, this process looks different only in its ontological focus—the *social* world as opposed to the *natural* world—from what we are pleased to call *scientific* progress. In the natural sciences as well, achieving adult competence is a matter of acquiring a complex family of perceptual, reflective, and behavioral skills in the relevant field. And there, too, such skills are embodied in an acquired set of structural, dynamical, and manipulational prototypes. The occasional deflationary voice to the contrary, our scientific progress over the centuries is a dramatic and encouraging reality, and it results in part from the myriad instructions (often painful) of an ongoing experimental and technological life lived under those same perception-shaping and behavior-guiding scientific prototypes.

Our conceptual development in the moral domain, I suggest, differs only in detail from our development in the scientific domain. We even have institutions whose job it is continually to fine-tune and occasionally to reshape our conceptions of proper conduct, permissible practice, and proscribed behavior. Civic, state, and federal legislative bodies spring immediately to mind, as does the civil service, and so do the several levels of the judiciary and their ever-evolving bodies of case-law and decision-guiding legal precedents. As with our institutions for empirical science, these socially focused institutions typically outlive the people who pass through their offices, often by centuries and sometimes by many centuries. And as with the payoff from our scientific institutions, the payoff here is the accumulation of unprecedented levels of recorded (social) experience, the equilibrating benefits of collective decision making, and the resulting achievement of levels of moral understanding that are unachievable by a single individual in a single lifetime.

To this overarching parallel it may be objected that science addresses the ultimate nature of a fixed, stable, and independent reality, while our social, legislative, and legal institutions address a plastic reality that is deeply dependent on the organizing activity of humans. But this presumptive contrast disappears almost entirely when one sees the acquisition of both scientific and moral wisdom as the acquisition of sets of *skills*. Both address a presumptively *im*plastic part of their respective domains—the basic laws of nature in the former case, and basic human nature in the latter. And both address a profoundly *plastic* part of their respective domains—the articulation, manipulation, and technological exploitation of the natural world in the case of working science, and the articulation, manipulation, and practical exploitation of human nature in the case of working morals and politics. A prosperous city represents simultaneous success in both dimensions of human cognitive activity. And the resulting artificial technologies, both natural and social, each make possible a deeper insight into the basic character of the natural universe, and of human nature, respectively.

Moral Unity/Systematicity

This parallel with natural science has a further dimension. Just as progress in science occasionally leads to welcome unifications within our understanding—as when all planetary motions come to be seen as special cases of projectile motion, and all optical phenomena come to be seen as special cases of electromagnetic waves—so also does progress in moral theory bring occasional attempts at conceptual unification—as when our assembled obligations and prohibitions are all presented (by Hobbes) as elements of a *social contract*, or (by Kant) as the local instantiations of a *categorical imperative*, or (by Rawls) as the reflection of *rules rationally chosen from behind a veil of personal ignorance*. These familiar suggestions, and others, are competing attempts to unify and systematize our scattered moral intuitions or antecedent moral understanding, and they bring with them (or hope to bring with them) the same sorts of virtues displayed by intertheoretic reductions in science, namely, greater simplicity in our assembled conceptions, greater consistency in their application, and an enhanced capacity (born of increased generality) for dealing with novel kinds of social and moral problems.

As with earlier aspects of moral cognition, this sort of large-scale cognitive achievement is also comprehensible in neurocomputational terms, and it seems to involve the very same sorts of neurodynamical changes as are (presumptively) involved when theoretical insights occur within the natural sciences. Specifically, a wide range of perceptual phenomena—which (let us suppose) used to activate a large handful of distinct moral prototypes, m_1, m_2, m_3, ..., m_n—come to be processed under a new regime of recurrent manipulation (recall the recurrent neuronal pathways of Figure 5.1b) that results in them all activating an unexpected moral prototype M, a prototype whose typical deployment has hitherto been in other perceptual domains entirely, a prototype that now emerges as a *superordinate* prototype of which the scattered lesser prototypes, m_1, m_2, m_3, ..., m_n can now be seen, retrospectively, as so many *sub*ordinate instances.

The preceding is a neural-network description of what happens when, for example, our scattered knowledge in some area gets *axiomatized*. But axiomatization, in the linguaformal guise typically displayed in textbooks, is but one minor instance of this much more general process, a process that embraces the many forms of *non*discursive knowledge as well, a process that embraces science and ethics alike.

Reflections on Some Recent "Virtue Ethics"

As most philosophers will perceive, the general portrait of moral knowledge that emerges from neural-network models of cognition is a portrait already under active examination within moral philosophy, quite independently of any connections it might have with cognitive neurobiology. Its original champion is Aristotle and its current research community includes figures as intellectually diverse as Mark Johnson (1993), Owen Flanagan (1991), and Alasdair MacIntyre (1981), all of whom came to this general perspective for reasons entirely of their own. For the many reasons outlined in the body of this essay, I am compelled (and honored) to count myself among them.

But I am not entirely comfortable in this group, for two of the philosophers just mentioned take a view, on the matter of moral progress, very different from that just outlined. Flanagan (1996) has expressed frank doubts that human moral consciousness ever makes much genuine "progress", and he suggests that its occasional changes are better seen as just a directionless meander made in local response to our changing economic and social environment.

MacIntyre (1981) voices a different but comparably skeptical view, wherein he hankers after the lost innocence of pre-Enlightenment human communities, which were much more tightly knit by a close fabric of shared social practices, which practices provided the sort of highly interactive and mutually dependent environment needed for the many moral virtues to develop and flourish. He positively laments the emergence of the post-Enlightenment, liberal, secular, and comparatively anonymous and independent social lives led by modern industrial humans, since the rich soil necessary for moral learning, he says, has thereby been impoverished. The familiar moral virtues must now be acquired, polished, and exercised in what is, comparatively, a social vacuum. If anything, in the last few centuries we have suffered a moral *regress*.

I disagree with both authors, and will close by outlining why. I begin with MacIntyre, and I begin by conceding his critique of the (British) Enlightenment's cartoon-like conception of *homo economicus*, a hedonic calculator almost completely free of any interest in or resources for evaluating the very desires that drive his calculations. I likewise concede his critique of the (Continental) Enlightenment's conception of *pure reason* as the key to identifying a unique set of behavior-guiding rules. And my concessions here are not reluctant. I agree wholeheartedly, with MacIntyre, that neither conception throws much light on the nature of moral virtue.

But as crude as these moral or meta-moral ideas were, they were still a step up from the even more cartoonlike conceptions of *homo sheepicus* and *homo infanticus* relentlessly advanced by the pre-Enlightenment Christian Church. Portraying humanity as sheep guided by a supernatural Shepherd, or as children beholden to a supernatural Father, was an even darker self-deception and was even less likely to serve as a means by which to climb the ladder of moral understanding.

I could be wrong in this blunt assessment, and if I am, so be it. For the claim of the preceding paragraph does *not* embody the truly important argument for moral progress at the hands of the Enlightenment. That argument lies elsewhere. It lies in the permanent opening of a tradition of cautious *tolerance* for a diversity of local communities each bonded by their own fabric of social practices; it lies in the establishment of lasting institutions for the principled *evaluation* of diverse modes of social organization, and for the institutionalized *criticism* of some and the systematic *emulation* of others. It lies, in sum, in the fact that the Enlightenment broke the hold of a calcified moral dictatorship and replaced it with a tradition that was finally prepared to learn from its deliberately broad experience and its inevitable mistakes in first-order moral policy.

Once again, I am appealing to a salient parallel. The virtue of the Enlightenment, in the moral sphere, was precisely the same virtue displayed in the scientific sphere, namely, the legitimation of responsible theoretical diversity and the establishment of lasting institutions for its critical evaluation and positive exploitation. It is this long-term process, rather than any particular moral theory or moral practice that might fleetingly engage its attention, that marks the primary achievement of the Enlightenment.

MacIntyre began his Introduction to *After Virtue* with a thought-provoking science fiction scenario about the loss of an intricate practical tradition that alone gives life to its corresponding family of theoretical terms, and the relative barrenness of their continued use in the absence of that sustaining tradition. This embodies the essentials of his critique of our moral history since the Enlightenment. But we can easily construct, for critical evaluation, a parallel critique of our *scientific* history since the same period, and that parallel, I suggest, throws some welcome light on MacIntyre's rather conservative perspective.

Consider the heyday of Aristotelian Science, from the fourth century BC to the seventeenth century AD (even longer than the Christian domination of the moral sphere), and consider the close-knit and unifying set of intellectual and technological practices that it sustained. There is the medical tradition running from Rome's Galen to the four Humors of the late medieval doctors. There is the astronomical/astrological tradition that extends through Alexandria's Ptolemy to Prague's Johannes Kepler, who was still casting horoscopes for the wealthy despite his apostate theorizing. There is the intricate set of industrial practices maintained by the alchemists from the Alexandrian Greeks to seventeenth-century Europe, which tradition simply owned the vital practices of metallurgy and metal-working, and of dye-making and medicinal manufacture as well. These three traditions, and others that space bids me pass over, were closely linked by daily practice as well as by conceptual ancestry, and they formed a consistent and coherent environment in which the practical and technological virtues of late antiquity could flourish. As they did. MacIntyre's first condition is met.

So is his second, for this close-knit "paradise" is well and truly lost, having been displaced by a hornet's nest of distinct sciences, sciences as diverse as astrophysics, molecular biology, anthropology, electrical engineering, solid-state physics, immunology, and thermodynamic meteorology. Modern science now addresses and advances on so many fronts that the research practice of individual scientists and the technological practice of individual engineers is increasingly isolated from all but the most immediate members of their local cognitive communities. And the cognitive virtues they display are similarly fragmented. They may even find it difficult to talk to each other.

You see where I am going. There may well be problems—real problems—arising from the unprecedented flourishing of the many modern sciences, but losing an earlier and somehow more healthy "golden age" is certainly not one of them. Though real, those problems are simply the price that humanity pays for growing up, and we already attempt to address them by way of interdisciplinary curricula, interdisciplinary conferences and anthologies, and by the never-ending search for explanatory unifications and intertheoretic reductions.

I propose, for MacIntyre's reflection, a parallel claim for our moral, political, and legal institutions since the Enlightenment. Undoubtedly there are problems emerging from the unprecedented flourishing of the many modern industrial societies and their sub-societies, but losing touch with a prior golden age is not obviously one of them. The very real problems posed by moral and political diversity are simply the price that humanity pays for growing up. And as in the case of the scattered sciences, we already attempt to address them by constant legislative tinkering, by the reality-driven evolution of precedents in the judicial record, by tolerating the occasional political "divorce" (e.g. Yugoslavia, the Soviet Union, the Scottish Parliament), and by the never-ending search for legal, political, and economic unifications. Next to the discovery of fire and the poly-doctrinal example of ancient Greece, the Enlightenment may be the best thing that ever happened to us.

The doctrinal analog of MacIntyre's implicit communitarianism in moral theory is a hyperbolic form of Kuhnian conservatism in the philosophy of science, a conservativism that values the (very real) virtues of any given "normal science" tradition (such as Ptolemaic astronomy, classical thermodynamics, or Newtonian mechanics) over the comparatively fragile institutions of collective evaluation, comparison, and criticism that might slowly force their hidden vices into the sunlight and pave the way for their rightful overthrow at the hands of even more promising modes of cognitive organization. One can certainly see Kuhn's basic "communitarian" point: stable scientific practices make many valuable things possible. But tolerant institutions for the evaluation and modification of those practices make even *more* valuable things possible—most obviously, new and more stable practices.

This particular defense of the Enlightenment also lays the foundation for my response to Flanagan's quite different form of skepticism. As I view matters from the neural-network perspective explained earlier in this essay, I can find no difference in the presumptive brain mechanisms and cognitive processes that underwrite moral cognition and scientific cognition. Nor can I find any significant differences in the respective social institutions that administer our unfolding scientific and moral consciousness respectively. In both cases, learning from experience is the perfectly normal outcome of both the neural and the social machinery. That means that moral progress is no less possible and no less likely than scientific progress. And since none of us, at this moment, is being shown the instruments of torture in the Vatican's basement, I suggest it is actual as well.

There remains the residual issue of whether the *sciences* make genuine progress, but that issue I leave for another time. The take-home claims of the present essay are that (1) whatever their ultimate status, moral and scientific cognition are on an *equal* footing, since they use the same neural mechanisms, show the same dynamical profile, and respond in both the short and the long term to similar empirical pressures; and (2) in both moral and scientific learning, the fundamental cognitive achievement is the acquisition of *skills*, as embodied in the finely tuned configuration of the brain's 10^{14} synaptic connections.

6 Reflections on the Neurobiology of Emotion and Feeling

Antonio R. Damasio

By the end of the nineteenth century Charles Darwin had made incisive observations on the expression of emotions in animals and humans and placed emotion in the perspective of biological evolution; William James had produced a scientific description of the phenomenon of emotion, thus opening the way to its experimental study; and Sigmund Freud was writing about the means by which emotion might play a role in psychopathology. Somebody freshly arrived on earth in 1994 and interested in the topic of emotion would have good cause to wonder why such groundbreaking developments did not lead to an assault on the neurobiology of emotion. What could possibly have gone wrong in the intervening century?

The simplest answer to this question is that emotion has received benign neglect from neuroscience and been passed over in favor of the study of attention, perception, memory, and language. The not-so-simple answer would invoke the following reasons. First, in spite of the auspicious beginnings provided by Darwin, James, and Freud, there was a manifest difficulty in defining the phenomenon of emotion. What was the nature and substance of emotion? What was its scope? What were its component manifestations? William James made a brave attempt as one can see in the following passage: "If we fancy some strong emotion and then try to abstract from our consciousness of it all the feelings of its bodily symptoms, we find we have nothing left behind, no 'mind-stuff' out of which the emotion can be constituted, and that a cold and neutral state of intellectual perception is all that remains" (James 1950). With these words, which were well ahead of both his time and ours, I believe William James actually seized upon the mechanism *essential* to the process of emotion and feeling. Uncharacteristically for him, the scope of the phenomenon implied in this definition, as well as the mechanisms and components presented in the remainder of his proposal, fell short of the complexity that needed to be addressed. As a result, James' statement has been the source of endless and sometimes hopeless controversy. (I cannot do justice here to the extensive scholarship on the controversy, but I recommend reviews by Paul Ekman (1993), Jerome Kagan (1989), Richard Lazarus (1991), George Mandler (1984), and Robert Zajonc (1984).) The issue with James was not just his stripping emotion down to a process which

Supported by a grant from the Mathers Foundation. This text is an adapted and updated version of A. R. Damasio 1995.

involved the body, shocking as that must have been to his critics and still seems to be to some contemporaries. The main problem was that James gave little importance to the process of evaluating mentally the situation which causes an emotion. His account works well for the first emotions one experiences in one's life, but not for many of the complex emotions we experience as adults. Equally problematic was the fact that, in the James view, the body is *always* interposed in the process of emotion. He made no provision for an alternative or supplementary mechanism to generate the feeling that corresponds to a body excited by emotion.

James postulated a basic mechanism in which particular stimuli in the environment excite, by means of an innately set and inflexible mechanism, a specific pattern of body reaction. In order for the reaction to occur there was no need to evaluate the significance of the stimuli. In many circumstances of our life as social beings, however, we know that emotions are triggered only after an evaluative, non-automatic, mental process. Because of the nature of our life experience, there is a broad range of stimuli and situations which have become associated with the stimuli which are innately set to cause emotions. The reaction to that broad range of stimuli and situations can be filtered by an interposed mindful evaluation, and because of the thoughtful, evaluative, filtering process, there is room for variation in the extent and intensity of pre-set emotional patterns, and there is thus a modulation of the basic machinery of the emotions so insightfully gleaned by James. Moreover, there seem to be other neural means to achieve the body sense which James considered the essence of the emotional process.

There is another reason for neurobiology's neglect of emotion. Although it cannot be said that defining the nature and scope of phenomena in perception, memory, or language has been an easy task, many investigators have found sufficient common ground for neurobiological research to proceed, on those topics, as fields of inquiry. The relative lack of agreement when it came to the matter of emotion simply perpetuated the idea that emotion was elusive and subjective, not quite ready to sit at the neurobiology table. This may have been the source for the bizarre distinction between cognition and emotion, as if somehow one could have thoughts without emotion, a mind without affect.

Finally, perhaps because an operational definition of the phenomenon has been so difficult to come by, the terminology used in studies of emotion has been problematic. The term *emotion* is often used to denote both the expressive, externalized aspects of this complex phenomenon and the internal experience that is only available to the subject who has the emotion. Some of the controversy surrounding the phenomenon actually results from confusion about which precise aspects are meant, in a given debate. As noted below, I believe that the use of the terms *emotion* and *feeling* to denote, respectively, the expressive and experiential aspects of the phenomenon, may be helpful in future investigations.

In spite of the neglect that emotion suffered during the twentieth century, some significant progress did take place. Here is a brief summary:

1.   A small number of cognitive scientists produced important analyses of the phenomenon of emotion, and characterized many of its components in rich detail (Ekman 1993; Kagan 1989; Lazarus 1991; Mandler 1984; Zajonc 1984; Schacter and Singer 1962).

2.   The processes of emotion were firmly linked to a set of phylogenetically older cortical regions and subcortical nuclei known as the "limbic system" (Bard 1928; Papez 1937; Hess 1954; Olds and Milner 1954; MacLean 1970). This important development was not without some problems of its own. The connection between limbic system and emotion is indeed valid, but it has aligned emotion with the older brain structures, and thus again separated it from the cognitive processes which have been aligned with the modern brain structures, namely the neocortex and its attending thalamic nuclei. In other words, the rift between emotion and cognition acquired a neuroanatomical counterpart in the duality between limbic system and neocortex.

3.   A number of human neuropsychological studies have demonstrated an important relation between human emotion and right hemisphere structures (Sperry 1981; Sperry, Gazzaniga, and Bogen 1969; Sperry, Zaidel, and Zaidel 1979; Bogen and Bogen 1969; Gardner *et al.* 1983; Borod 1992; R. Davidson 1992; Heilman, Watson, and Bowers 1983; Gainotti 1972; Rapcsak, Comer, and Rubens 1993); and, in spite of the shortcomings of the link between emotion and limbic system, the perceived relation did bring to focus limbic structures such as the amygdala and the anterior cingulate cortex. Recent studies in primates, rodents, and humans have shown beyond the shadow of a doubt that the amygdala plays a pivotal role in emotion (Aggleton and Passingham 1981; Rolls 1992; Davis 1992; LeDoux 1992; LeDoux 1993; Halgren 1992; Adolphs and A. R. Damasio 1998).

Other studies have pointed out that the anterior cingulate cortex is a likely component of the circuitry required for emotional processing (A. R. Damasio and Van Hoesen 1983; A. R. Damasio 1994), and there is new evidence to suggest that specific subsets of the prefrontal cortices, of somatosensory cortices, and of some higher-order cortices in right hemisphere are specifically related to the processes of emotion (A. R. Damasio 1994, 1996). Moreover, the role of basal forebrain, hypothalamus, and brain-stem in the process of emotion has been established beyond the shadow of a doubt (see A. R. Damasio 1999 for review; also A. R. Damasio *et al.* 1998, Panksepp 1991).

When the above evidence is considered together with findings from other fields of neurobiology (e.g. neurophysiology, neurochemistry and neuropharmacology, and neuroendocrinology), it is possible to formulate some operational concepts and a hypothesis that might be potentially useful in future research on the neurobiology of the emotions. This is what I present in the next few paragraphs, noting that, given the limited space of this chapter, the discussion is centered on the level of large-scale neural systems, although the implications of the hypothesis extend to both the cognitive level and the operation of neural structures at smaller scale.

What is an Emotion?

The word *emotion* usually evokes one of the so-called "primary" or "universal" emotions—happiness, sadness, fear, anger, surprise, or disgust—but it is important to note that there are numerous other behaviors to which the label emotion has been attached. They include so-called "secondary" or "social" emotions, such as embarrassment, jealousy, guilt, or pride, and what I call "background emotions", such as well-being or malaise, calm or tension, anticipation and exploration. The label of emotion has also been attached to drives and motivations, and to the states of pain and pleasure. Deciding on what constitutes an emotion is not easy, and when one surveys the whole range of possible phenomena, one does wonder if any sensible definition of emotion can be formulated, and if a single term remains useful to describe all these states. Others have struggled with the same problem and concluded that it is hopeless. At this point, however, my preference is to retain the traditional nomenclature, clarify the use of the terms, and wait until further evidence dictates a new classification, my hope being that by maintaining some continuity we will facilitate communication at this transitional stage. I will talk about three levels of emotion—background, primary, and secondary. I see drives and motivations, and pain and pleasure, as triggers or constituents of emotions, but not as emotions in the proper sense.

All of the above phenomena share a biological core which can be outlined as follows:

1.   Emotions are complicated collections of chemical and neural responses, forming a pattern; all emotions have some kind of regulatory role to play, leading in one way or another to the creation of circumstances advantageous to the organism exhibiting the phenomenon; emotions are *about* the life of an organism, its body to be precise, and their role is to assist the organism in maintaining life.

2.   Notwithstanding the reality that learning and culture alter the expression of emotions and give emotions new meanings, emotions are biologically determined processes, depending on innately set brain devices, laid down by a long evolutionary history.

3.   The devices which produce emotions occupy a fairly restricted ensemble of subcortical regions, beginning at the level of brain stem and moving up to the higher brain; the devices are part of a set of structures which both regulate and represent body states.

4.   All the devices can be engaged automatically, without conscious deliberation; the considerable amount of individual variation, and the fact that culture plays a role in shaping some inducers, does not deny the fundamental "stereotypicity", automaticity, and regulatory purpose of the emotions.

5.   All emotions use the body as their theater (internal milieu, viscera, vestibular and musculoskeletal systems), but emotions also affect the mode of operation of numerous brain circuits: the variety of the emotional responses is responsible for profound changes in both the body landscape and the brain landscape. The collection of these changes constitutes the substrate for the neural patterns which eventually become feelings of emotion.

In short, the core of an emotion is a collection of changes in body state and brain state, induced in myriad organs and in some brain circuits, under the control of a dedicated brain system which is responding to the content of one's thoughts relative to a particular entity or event. The responses toward the body proper result in a specific body state, and those toward the brain itself result in a specific mode of network operation which entails a change in cognitive style. The former produce physiologic modifications, many of which are perceptible to an external observer, for example, changes in skin color, body posture, and facial expression. The latter produce changes that are mostly perceptible to the individual in whom they are enacted.

The core of the phenomenon of emotion is completed by (i) a mental evaluative process, which precedes the "dispositional responses" outlined above, and (ii) the subject's perception of the changes induced by the responses, for which I reserve the term feeling. The full range of the phenomenon of emotion, in its most traditional sense, thus includes (1) evaluation, (2) dispositions to respond, and (3) feeling, although the feeling component is of such importance in the understanding and future investigation of the phenomenon that it deserves the separate definition presented below.

What is a Feeling?

As body and brain change under the influence of dispositions to respond, the subject can perceive those changes and monitor their unfolding. What I call a *feeling* is (*a*) a mental image of our body state while thoughts about specific contents are processed, *and* (*b*) the mental imaging of the cognitive mode with which such contents are processed (e.g. the rate at which images are generated, how intensely they are attended, how they participate in inferential processes and choice of actions, and so on). If an emotion is a collection of body state and brain state changes connected to particular mental images whose evaluation activated a specific brain system, *feeling an emotion is the imaging of those changes in juxtaposition to the mental images which initiated the cycle.*

The body state changes specific to emotions are enacted by neural signals (e.g. autonomic, musculoskeletal), and chemical signals (e.g. endocrine). The brain state changes are enacted by neural signals towards neurotransmitter and neuromodulator nuclei in hypothalamus, brain stem and basal forebrain, which in turn send signals to a variety of neural sites, such as basal ganglia, cerebral cortex. Direct chemical signalling from the body proper also affects the operation of brain networks. The changes in cognitive mode I mentioned above are the result of these brain state changes (A. R. Damasio 1994, 1999).

A Neurobiological Hypothesis

To explain the working hypothesis which currently guides the work in my laboratory, let me ask you to consider an example drawn from your adult experience, for instance, the meeting of a friend whom you have not seen in a long time or the news of the unexpected death of a person who worked closely with you. I hypothesize the following:

Stage 1.   The process begins with the conscious and deliberate considerations you entertain about the situation. These considerations are expressed as mental images organized in thoughts, and they concern myriad aspects of your relationship with, say, a given person, reflections on the current situation and its consequences for you, and others; in short, a cognitive evaluation. Some of the images you conjure up are non-verbal (e.g. the likeness of a person or place), others are verbal (e.g. words and sentences which comment on the person's attributes, or a person's name, etc.). The neural substrate for such images is a collection of several widely distributed, topographically organized representations, occurring in varied early sensory cortices (e.g. visual, auditory). These representations, in turn, are constructed under the guidance of dispositional representations held in distributed manner over a large number of higher order association cortices (A. R. Damasio and H. Damasio 1994).

Stage 2.   Networks in the prefrontal cortex respond automatically, involuntarily, non-consciously, to signals arising from the processing of the imagery described above. This prefrontal response comes in part from dispositions which embody knowledge pertaining to how *certain types of situation* have usually been paired with *emotional responses*, in your *individual experience*. In other words, it comes from both *acquired* and *innate* dispositions, the acquired dispositions having been obtained under the influence of those which were innate. The acquired dispositions embody the *unique* experience you have had of such relations in your life.

In brief, the prefrontal dispositions are separate from the innate dispositions needed for primary emotions (e.g. happiness, sadness, fear, anger,
disgust, surprise) which are based in brain stem nuclei. But acquired dispositions (see Stage 3 below) require innate dispositions in order to express
themselves.

Stage 3.   Automatically, involuntarily, and "non-consciously", the responses
of the prefrontal dispositions described in Stage 2 are signalled to structures
such as amygdala, basal forebrain, hypothalamus, and brain stem nuclei. In
turn, dispositional representations in the latter regions respond in the following ways: (*a*) by activating autonomic nervous system nuclei and signalling to
the body, via peripheral nerves, with the consequence that viscera are placed
in the state that is most commonly associated with the type of triggering
situation; (*b*) by dispatching signals to the motor system such that the skeletal
muscles could complete the external picture of an emotion in facial and body
posture expressions; (*c*) by activating endocrine and peptide systems whose
chemical actions also result in body state changes and brain state changes (the
latter of which induce specific behaviors and alter the mode in which body
signals are conveyed to the brain); and (*d*) by activating non-specific neurotransmitter nuclei in brain stem and basal forebrain, which, in turn, release
their chemical messages (in the form of dopamine, epinephrine, serotonin,
and acetylcholine release) in varied regions of the telencephalon (e.g. basal
ganglia, cerebral cortex), and alter their mode of processing.

The changes caused by (*a*), (*b*) and (*c*) impinge on the body, cause an
"emotional body state", and are subsequently signalled back to the brain, not
only to the limbic system but also to the somatosensory system. The changes
caused by (*c*) and (*d*), which do not arise in the body proper but rather in
body-related brain stem and basal forebrain structures, modify both the mode
of body signalling and the style and efficiency of cognitive processes.

The neural basis for feelings includes both sets of representation, that is,
those which correspond to body changes and those that correspond to
changes in network operation. All of these representations are instantiated
on-line, in sensory maps of the brain stem nuclei (such as the parabrachial
nucleus), of the hypothalamus, and of somatosensory cortices such as insula,
SII, and SI. It is noteworthy that those representations constitute the substrate
of feelings but they are not sufficient for "knowledge" of feelings to occur.
The additional mechanisms which allow us to know a feeling are part of
the complex operations of consciousness. They are outside the scope of this
review and are discussed in detail in A. R. Damasio 1999.

The part of the hypothesis which deals with responses aimed at the body is
an elaboration of the James's original proposal. To the neural route of signal-
ing implicit in William James, I have added a chemical route, from brain to
body-proper and from body-proper back to brain. The part of the hypothesis
which concerns responses aimed at nuclei in the brain core is new, and so is

the notion that subsequent responses from those nuclei, linked to the process of emotion, alter the activity of networks which support cognitive processes. There is a considerable amount of evidence in favor of this formulation available in studies of experimental neuroanatomy, neurophysiology, neuroendocrinology, neuropsychology, and general biology (see A. R. Damasio 1999 for review).

One of the criticisms leveled at William James concerns his idea that we must use the body to enact emotions. Although I believe that in many situations emotions and feelings operate precisely in that manner, from mind/brain to body, and back to mind/brain, I also believe that in numerous instances the brain learns to generate the fainter image of an "emotional body state", without having to enact it in the body-proper. Moreover, the activation of neuromodulatory nuclei in brain stem and basal forebrain, and their responses to other brain sites, entirely bypasses the body, although, in a curious way, these nuclei are part and parcel of the neural representation of body regulation.

There are thus neural devices which allow us to feel "as-if" we were having an emotional state, as if the body were being activated and modified. Such devices permit us to bypass the body and avoid a slow and energy-consuming process. We conjure up some semblance of a feeling within the brain alone.

"As-if-body" devices would have been developed as we were growing up and adapting to our environment. The association between a certain mental image and the surrogate of a body state would have been acquired by repeatedly associating the images of given entities or situations with the images of freshly enacted body states. To have a particular image trigger the bypass device, it was first necessary to run the process through the body theater, to loop it through the body, as it were.

Several controversies which have surrounded studies on emotion, and James's proposal in particular, may be resolved by this hypothesis. For instance, I propose that the body is only necessarily interposed in the process of emotion during development. Following development the body may or may not be activated in emotional processing, although patterns of neural representation of body state will still be engendered within the brain ("as if" emotional body states). The body relatedness of the process of emotion and feeling is never lost in "as-if-body" states but the body itself can be bypassed.

Feelings are as Cognitive as any other Percept

In the perspective of the hypothesis outlined above, feelings are just as cognitive as any other percept. It is true that feelings are about *something different* from most other images. But what makes them different is that they are first and foremost about the body, that they offer us *the cognition of our internal milieu, visceral, and musculoskeletal states*, as they become affected by pre-organized mechanisms and by the cognitive structures we have developed under their influence. Feelings are the first step in letting us *mind the body*. They let us mind the body directly, when they give us images of the body, or indirectly, as when they give us recalled images of the body state appropriate to certain situations.

Feelings give us a glimpse of what goes on in our flesh, as a momentary image of that flesh gets juxtaposed to the image of other objects and situations, and in so doing, modify the comprehensive notion we get of those other objects and situations. By dint of juxtaposition body images give to other images a *quality* of goodness or badness, of pleasure or pain.

What are Emotions and Feelings for?

There are several possible answers to this question, but the one I prefer states that emotion is a critical, albeit indirect, element in biological regulation rather than being a luxury. Emotions and feelings are closely linked to the behaviors necessary for survival. Feeding, the avoidance of danger, the seeking of advantageous physical and social conditions, and reproduction are not possible without the complex signals that are provided to the individual and to others by feelings and emotions. In general terms, then, emotion and feeling help achieve homeostasis, albeit indirectly, and assist, in so doing, with communication among individuals of the same and other species. Furthermore, as I have recently proposed, emotions and feelings may play an important role in the processes of reasoning and decision-making, especially those that relate directly to person and immediate social environment (A. R. Damasio 1994; see also De Sousa 1991; Johnson-Laird and Oatley 1992).

By itself, the emotional response can accomplish several useful goals (e.g. rapid concealment from a predator, a frightening display of anger towards a competitor). However, the process does not stop with the bodily changes which define an emotion. In humans, at least, the cycle continues. The next step is the *feeling of the emotion* in connection to the entity that excited it, and that step is followed by the *knowing of the feeling*, the realization that there is a link between entity and emotional body state. One may well ask why should one become cognizant of such a relation and complicate matters by bringing consciousness into this process, given that one already has a means to respond adaptively at automated level. The answer is that consciousness buys us an enlarged protection policy. If we come to *know* that animal X causes fear, we will now have two ways of behaving relative to the animal. The first way is innate: we do not control it, and it is not specific to animal X; many creatures and situations can cause the same response. The second way is based on our own experience and is specific to animal X. Knowing about animal X allows us to think ahead and predict the probability of X being present in a given environment so that we can preemptively *avoid* X, rather than just have to react to X's presence in an emergency. But there are other advantages of feeling one's emotional reactions and knowing of those feelings. For instance, one can generalize knowledge, and decide to be cautious with any object that looks *like X*. Feeling an emotion offers us flexibility of response based on the particular history of our interactions with the environment.

The key components of this hypothesis are being tested in humans, in both neurological patients with lesions and normal individuals participating in imaging experiments. Some aspects of the hypothesis are also being tested in experimental animals.

Emotions and feelings are not intangible phenomena. Their subject matter is concrete and they can be related to specific systems in body and brain, no less so than vision or speech. Nor are the responsible brain systems confined to the subcortical sector. Brain core and cerebral cortex work together to construct emotion and feeling, no less so than they do in vision.

7 Words and Concepts in the Brain

Hanna Damasio

Introduction

The goal of my work is the understanding of the neural basis for certain cognitive processes and for certain behaviors. The level at which I work is the large-scale system level in the primate brain, specifically, in the human brain. I am interested in events at other levels, as for instance at cellular level or at the level of microcircuitry, or at the purely cognitive level, all of which are obviously necessary to get to a complete understanding of mind and brain phenomena, but my work is about the large-scale system level and this is the subject of this chapter.

Work in this general area has a long tradition but the current approaches and methods are actually quite new.

Functional imaging, for instance, as practiced today, is literally just beginning as an approach. We can use it to perform experiments in normal individuals as well as in neurological patients. We take advantage of either the regional distribution of a radiolabled tracer in the human brain; or we make use of the effects of a changing magnetic field on brain physiology. In either case, we measure an index of brain function *when* the brain is involved in performing a particular cognitive behavioral task.

With the modern *lesion method*, which remains the gold standard for bridging brain and cognition, focal areas of brain damage, which we call lesions, are used as probes to test hypotheses about the possible function of a system. The system in question is always made up of several components, and the function is always the result of the combined operation of the components within the system. The lesion is a "knock-out" of one component, and it allows us to test our prediction as to what should happen when the component is knocked out. In fact, it is the same strategy used in molecular biology when you selectively knock out a gene and verify if the results of the lack of that gene conform to your prediction.

Although lesion work is really the oldest in neurosciences, the systematic lesion method approach is quite recent. It could only appear after modern neuroimaging techniques allowed us to study the human brain *in vivo* (see H. Damasio 1995). Just as with functional imaging, and perhaps even more so, proper lesion method experiments require a complicated set up of human and technical resources. This probably explains why it is so rarely practiced.

Supported by NINDS Program Project Grant NS 19632. Parts of the data reported here have appeared in the following publications: H. Damasio *et al.* 1996; Tranel, H. Damasio, and A. R. Damasio 1997; Grabowski, H. Damasio, and A. R. Damasio 1998.

What I would like to report here is the result of a number of studies on the neural correlates of language, specifically, the neural systems that are possibly involved in word retrieval, and in the corresponding concept retrieval.

If you open a textbook of neurology, neuropsychology, or linguistics, in a section concerning the possible neural basis of language, you are likely to read that there is an anterior region in the left hemisphere, more precisely in the left frontal operculum, that allows the production of words, and a posterior region, in the left superior temporal gyrus, which allows the comprehension of the spoken word. These two areas are named for their "discoverers" as Broca's area and Wernicke's area, and they have been with us since the second half of the nineteenth century (Broca 1861; Wernicke 1874). In the 1960s Norman Geschwind (1965) revived the interest in these areas and added a few others, also in the left hemisphere, that he believed were involved in the production and comprehension of language. These included the inferior parietal lobule—parts of the gray matter in the supramarginal and angular gyrus, as well as the white matter underneath, the famous arcuate fasciculus connecting Wernicke's and Broca's areas. He also added part of Brodmann's area 37 to the language map. Such a diagrammatic representation was fine at the time; but, unfortunately, it is still the prevailing anatomical framework for the discussion of the neural basis of language. We all know, of course, that the story is not this simple. The problem with the old anatomical account is not that it is wrong but that it is quite incomplete.

It is still useful to maintain such designations as Broca's aphasia and Wernicke's aphasia, in the sense that they predict likely loci of brain damage in a neurological patient. The designations are clinically useful. But we no longer accept the idea that there are only these two language-related areas, connected by a direct and unidirectional pathway that translates thought into words. The way we see the production of language today involves many brain regions connected bidirectionally, forming complex systems.

Where does the idea that there are language areas beyond Broca and Wernicke come from? As usual, individual observations gave us the first clues, long before systematic experiments could demonstrate it. For instance, we have solid evidence that a subject with damage to the anterior sector of the left temporal lobe, involving the temporal pole, in spite of perfectly fluent, non-aphasic language, will be severely impaired in the retrieval of names for specific persons. The patient knows the person and can provide verbal descriptions that allows an independent examiner, who does not know what stimulus the patient is looking at, to identify the person the patient is trying to name. On the other hand such a patient can name perfectly well other objects, either natural or man-made, and can also look at pictures depicting actions and retrieve the correct word denoting those actions. So what we find is a very circumscribed deficit of word retrieval and a circumscribed area of brain damage in the left temporal pole, away from either Broca's area or Wernicke's area, which are not damaged at all.

On the other hand a subject with a lesion in the left temporal lobe, but located posteriorly in the temporo-occipital junction away from the temporal pole, as well as from Wernicke's and Broca's area, will show a deficit in the retrieval of words denoting manipulable objects. The retrieval of words for persons or animals is entirely normal.

Many observations such as these led us to construe the following hypotheses:

General hypothesis. The retrieval of words which denote concrete entities belonging to distinct conceptual categories depends upon relatively separable regions in higher-order cortices of the left temporal lobe.

Specific hypothesis. The retrieval of words which denote concrete entities belonging to distinct conceptual categories depends upon relatively separable regions in higher-order cortices of the left temporal lobe.

In order to address these hypotheses we studied a large group of subjects with single lesions of either hemisphere so that we could sample vast sectors of the telencephalon.

1. Lesion Studies

Methods

Subjects Subjects with unilateral brain damage ($n = 116$) were selected from the Patient Registry of the University of Iowa's Division of Cognitive Neuroscience. All gave informed consent in accordance with the Human Subjects Committee of the University of Iowa. As a group, the subjects permitted us to probe most of the telencephalon—their lesions were located in the left ($n = 68$) or right ($n = 48$) hemisphere, in varied regions of the cerebral cortex. Lesions were caused by either cerebrovascular disease ($n = 95$), herpes simplex encephalitis ($n = 5$), or temporal lobectomy ($n = 16$). Handedness, measured with the Geschwind–Oldfield Questionnaire which has a scale ranging from full right-handedness ($+100$) to full left-handedness (-100), was distributed as follows: 109 were right-handed ($+90$ or greater); 4 were left-handed (-90 or lower); 3 were mixed-handed ($< +90$ and > -90).

All subjects had been extensively characterized neuropsychologically and neuroanatomically, according to the standard protocols of the Benton Neuropsychology Laboratory (Tranel 1996) and the Laboratory of Neuroimaging and Human Neuroanatomy (H. Damasio and A. R. Damasio 1989; H. Damasio and Frank 1992; H. Damasio 1995). All subjects had at least average verbal intelligence (as measured by the WAIS-R), a high school education or higher, and no difficulty attending and perceiving visual stimuli. Some subjects in the sample were recovered aphasics, but in no case was the residual aphasia of such severity as to preclude the production of scorable recognition responses. Subjects with severe aphasia or with severe visual perceptual deficits were not included in the brain-damaged sample.

We studied 55 normal controls, who were matched to the brain-damaged subjects on age, education, and gender distribution.

Stimuli The unique stimuli were pictures of persons (presented as faces), drawn from the Iowa Famous Faces Test ($n = 77$) (Tranel, H. Damasio, and A. R. Damasio 1995) and a modified version of the Boston Famous Faces Test ($n = 56$; Albert, Butter and Levin 1979). The faces used in the experiments were of persons who would have been known to the subjects prior to the onset of their brain damage. Face pictures were prepared as black-and-white slides in which all non-face background features were deleted. The non-unique stimuli were pictures of animals ($n = 90$) and tools ($n = 104$) drawn from the Snodgrass and Vanderwart (1980) line drawings and from a set of photographs we have prepared (A. R. Damasio *et al.* 1990). The categories of animals and tools did not differ in word frequency, as derived from the Francis and Kucera (1982) database.

The choice of the three types of item was dictated by previous studies and by the fact that the entities represent varied conceptual categories and have a diverse characterization in terms of sensory and motor specifications and contextual complexity (A. R. Damasio and H. Damasio 1994). These entities vary along a number of dimensions which characterize their properties and the interactions of the individual with the entities, including sensory channels, used in the interaction: motor pattern (if applicable—for instance, manipulability); natural or artifactual status; physical characteristics; capability for self-movement; and so on (see Tranel, H. Damasio, and A. R. Damasio 1997 for a detailed explanation of these factors). 'Contextual complexity' refers to the complexity of relationships which help constitute an item, along with its physical characteristics. Contextual complexity permits us to classify a given concrete entity along a dimension that ranges from *unique* (an entity belonging to a class with $N = 1$ and depending on a highly complex context for its definition), to varied *non-unique* levels (entities processed as belonging to classes with $N > 1$, having many members whose definition depends on less complex contexts).

Procedure The stimuli were shown in random order one-by-one on a Caramate 4000 slide projector, in free field. For each stimulus, the subject was asked to tell the experimenter what (or who) the entity was. If the subject produced a vague or superordinate-level response (e.g. "something you can work with" or "a movie star"), the subject was prompted to "be more specific; tell me exactly what [who] you think that [thing] is." Prompting was repeated if the experimenter sensed that the subject could generate a more specific answer, or if the subject produced a paraphasic response that might be difficult to score. Time limits were not imposed. All responses were audiotaped and prepared as typewritten transcriptions. The transcriptions were scored by raters who were blind to the experimental hypotheses, following the procedures specified below; when necessary, the raters also used information from the audiotapes.

Neuropsychological Data Quantification and Analysis   The dependent measure was a recognition score and a naming score. For each stimulus the subject's recognition/naming response was scored as follows. First, if the stimulus was named correctly, the item was scored as a correct recognition and naming. In other words, we accepted correct naming as unequivocal evidence of correct recognition. Our rationale for this approach is that we have never found a subject who would produce a correct name, and then fail to recognize the stimulus that was named. As part of one of our studies, we explored in a subset of subjects with correct naming responses whether they had retrieved the concept for an item prior to retrieving its name, and, as we expected, they had. In fact, we do not believe it is possible for someone to name, accurately and reliably, an unrecognized item, even in the extreme instance of patients with Alzheimer's disease who may seem, on the face of it, to do just that. (A severely inattentive or demented patient may produce a naming response and then, by the time the response comes under scrutiny, have lost from working memory the material recalled during concept retrieval, leading to the appearance that the patient has named but not recognized. We believe this is an artifact of the attentional/memory defect, though, and we stand by the claim that concept retrieval remains a prerequisite for accurate naming.)

Second, for items that were not named correctly, the subject's responses were presented (as typewritten transcriptions) to raters who were asked to determine what the stimulus was *from the description alone*, without having in front of them either the stimulus or its name. Thus, if the subject had provided a specific description of the entity, including information about characteristics, functions, and properties (e.g. "that's the president who was killed . . . his brother was also killed, later . . . he had an affair with that movie star who killed herself"; or "that's an animal that you find on farms; it makes an oinking sound and likes to roll in the mud"), that was sufficiently detailed so that raters could identify the entity from the description alone, without having in front of them the stimulus or its name, the response was scored as correct recognition but as a failure in naming. If on the other hand it was not possible for the rater to come up with the correct name of the stimulus based on the response of the subject, the item was classified as a failure in recognition which precluded correct naming. Therefore the item was not counted as a failure in naming.

For each subject and each category, the number of correct recognition responses was divided by the number of stimuli in the category and multiplied by 100 to yield a percent correct recognition score. The naming score was calculated by summing the number of correct naming responses using *only* those stimuli for which the subject had produced a correct recognition response. Classification of brain-damaged subjects as normal or abnormal on the six tasks was conducted by calculating for each subject the extent to which their scores differed from the means of the normal controls.

Neuroanatomical Data Quantification and Analysis   The neuroanatomical analysis was based on magnetic resonance (MR) data obtained in a 1.5 Tesla scanner with an Spg sequence of thin (1.5 mm) and contiguous T_1 weighted coronal cuts, or in those subjects in whom an MR could not be obtained on computerized axial tomography (CT) data. All neuroimaging data were obtained in the chronic epoch (at least three months post-onset of lesion). Each subject's lesion was reconstructed in three dimensions using Brainvox (H. Damasio and Frank 1992; Frank, H. Damasio, and Grabowski 1997). The anatomical description of the lesion overlap and of its placement relative to neuroanatomical landmarks was performed with Brainvox, using the MAP-3 technique. All lesions in this set were transposed and manually warped into a normal 3-D brain, so as to permit the determination of the maximal overlap of lesions relative to subjects grouped by neuropsychological defect. A detailed description of MAP-3 is provided in Frank, H. Damasio and Grabowski 1997; in brief, it entails: (1) a normal 3-D brain (the template brain) is resliced so as to match the slices of the MR/CT of the subject and create a correspondence between each of the subjects' MR/CT slices and the normal resliced brain; (2) the contour of the lesion on each slice is then transposed onto the matched slices of the normal brain, taking into consideration the appropriate anatomical landmarks; (3) for each lesion the collection of contours constitutes an "object" that can be co-rendered with the normal brain. The objects in any given group can intersect in space, and thus yield a maximal overlap relative to both surface damage and depth extension. The number of subjects contributing to this overlap is thus known.

Results

Impaired Word Retrieval   The subjects were first classified according to their task performance, which was normal (within 2 standard deviations of the mean of the normal controls) in 97 subjects (47 with left hemisphere damage, 50 with right) and abnormal (2 or more standard deviations below normal) in 19. All but 1 of the latter had damage in left hemisphere, thus supporting the prediction that disruption of word retrieval for all three categories would be correlated with lesions in left but not right hemisphere. The scores of brain-damaged subjects who performed abnormally in each of the three word categories were significantly lower (all P's <0.001) than those of the brain-damaged subjects who performed normally.

In addition to isolated word retrieval defects for unique persons, for animals, or for tools, the instances of two combined defects always involved "persons/animals" or "animals/tools", but never "persons/tools". When naming of persons and tools was impaired in the same individual, naming of animals was impaired as well. The nonoccurrence of instances of the persons/tools combination is statistically significant ($P<0.001$). The anatomical evidence that follows shows why this combination of defects is not possible on the basis of a single lesion.

The lesions of the 18 left hemisphere subjects with abnormal performance revealed the following: maximal overlap of lesions for abnormal retrieval of words for persons was found in left temporal pole (TP); maximal overlap of lesions for abnormal retrieval of words for animals was found in left inferotemporal cortex (IT; mostly anterior); maximal overlap of lesions for abnormal retrieval of words for tools was found in posterolateral IT, along the junction of lateral temporo-occipito-parietal cortices (a region we have designated as "posterior IT+"). In all instances the lesion overlap encompassed both cortical and subcortical white matter. Thus the prediction that the disruption of word retrieval for each category would be correlated with partially separable neural sites within left higher-order cortices of the temporal region was supported, qualified only by noting that impaired retrieval of words for tools was correlated with damage that extended into left inferior parietal cortex and the temporo-occipital junction. To assess the reliability of the findings, we conducted a test using anatomical placement of lesion as the independent variable, and compared, for each word category, the naming scores of three groups of subjects: those with lesions in TP only; those with lesions in posterior IT+ only; and those with lesions centered in IT (whose outer borders trespassed, in some instances, into either TP or posterior IT+, but not into both). A MANOVA yielded highly significant results for persons ($F(2,20) = 14.85$, $P<0.001$) and tools ($F(2,20) = 6.65$, $P<0.01$), and a marginally significant result for animals ($F(2,20) = 3.30$, $P = 0.058$), supporting the conclusion that the anatomical placement of lesion is a crucial factor determining word retrieval performance. The close link between neuroanatomical structure and lexical category is also evident at individual subject level.

In sum, impaired retrieval of words for concrete entities correlated with damage in higher-order cortices outside classic language areas; moreover, impairments in the three word categories were correlated in a consistent manner with separable neural sites. Two of the latter, TP and posterior IT+, are not even contiguous, do not overlap cortically or subcortically, and are so distant as to make it virtually impossible for a single lesion to compromise them without also compromising the intervening region. This explains why a combined defect for persons and tools never occurred in our sample.

Impaired Retrieval of Concepts   What if instead of looking at the maximal overlap of lesions in relation to defective word retrieval we looked at maximal overlap of lesions relative to defective recognition of the presented item, that is, defective concept retrieval? I should mention here that all subjects are tested for strictly visual perceptual defects with a large battery of psychophysical experiments and that none of the subjects in this group has a perceptual visual defect that could explain the failure to recognize a visually presented stimulus.

The subjects were first classified according to their task performance, which was normal (within 2 standard deviations of the mean of the normal controls) in 76 subjects (27 with right hemisphere damage, 49 with left) and abnormal (2 or more standard deviations below normal) in 40 (21 with right hemisphere damage, 19 with left). The scores of brain-damaged subjects who performed abnormally in each of the three conceptual categories were significantly lower (all $Ps < 0.0001$) than those of the brain-damaged subjects who performed normally.

A number of subjects had concept retrieval defects for one single category: persons ($n = 7$), animals ($n = 19$), or tools ($n = 5$). There were also subjects in whom concept retrieval was affected in two categories. The combination "person/animals" was found in 6 subjects; the combination "animals/tools" was found in 3 subjects. However, no subject showed a combined "person/tools" defect, and in no instance were all three categories involved. In order to test the reliability of the negative findings concerning the persons/tools and persons/animals/tools combinations we conducted a Fisher Exact Probability Test on the following matrix (where n is the number of subjects with the designated set of defects): persons and tools ($n = 0$), persons but not tools ($n = 13$), tools but not persons ($n = 8$), neither tools nor persons ($n = 19$). The test yielded a P-value of 0.029, supporting the conclusion that the nonoccurrences of combined persons/tools and persons/animals/tools defects were not due to chance. Once again, the anatomical evidence given next will show why these combinations cannot be found with single unilateral lesions.

The overlap of lesions of the 40 subjects with abnormal performance revealed that: (*a*) in subjects with abnormal retrieval of concepts for *persons*, lesion overlap was maximal in right temporal polar region; (*b*) in subjects with abnormal retrieval of concepts for *animals*, lesion overlap was maximal (i) in right mesial occipital (mostly infracalcarine), extending into mesial ventral temporal region, and (ii) in a second locus (with a smaller number of overlaps) in the left mesial occipital region; (*c*) in subjects with abnormal retrieval of concepts for *tools*, lesion overlap was maximal in left lateral occipital—temporal—parietal junction. In all cases, the maximal overlaps concerned both cortex and subcortical white matter, as was seen for abnormal word retrieval.

In sum, impaired retrieval of concepts for concrete entities correlated principally with damage in higher-order cortices in right temporal polar and mesial occipital/ventral temporal regions, and in lateral occipital-temporal-parietal regions. The spatial distribution of the lesion loci critical for the appearance of these defects explains why a defect involving persons and tools could not be found in subjects with unilateral lesions. The loci for those two categories are in different hemispheres. Hence, only a suitable bilateral lesion is likely to produce this combination. The same applies to an impairment involving persons, animals, and tools.

2. PET Studies

What happens in normal subjects when they perform the same tasks used in the lesion studies? Are the areas revealed by the lesion overlaps particularly engaged during the performance of comparable word retrieval tasks? In order to answer this question we have been performing PET studies in normal individuals, using subsets of the same stimuli and precisely the same paradigm. The hypotheses remain the same:

General hypothesis. Access of word forms for visually presented concrete entities depends on activation of the left inferior temporal cortex and/or the left temporal pole.

Specific hypothesis. Accessing proper nouns will activate the left temporal pole; accessing common nouns the left IT. Activation patterns in IT will be different for retrieval of words denoting animals and tools/utensils.

Methods

Stimuli   We used stimuli from the test batteries employed in the lesion study for the target tasks and a set of unknown human faces presented either right-side up or upside down for the control task. They were presented in black and white on a video screen suspended 15 inches from the subject, using a microcomputer-driven video laser disk. In a pilot study with normal subjects, we determined empirically the rate of presentation for each set of stimuli that would produce error rates of approximately 10 percent. In order to achieve similar levels of performance, the rate of stimulus presentation had to be different among the tasks. Familiar faces were presented every 2.5 sec., tools every 1.8 sec., and animals every 1.5 sec. (stimuli for the control task were presented every 1.0 sec.). Each stimulus was on screen for 25 percent of the stimulus time cycle; otherwise the screen was black. Faces for the person naming task were selected during a pilot session 24–48 hours before PET by having the subjects view a collection of famous faces from the Iowa and Boston Famous Faces tests and indicate whether they recognized each person. During the pilot session, subjects were told that they should not name any of the persons, and no names were provided by the experimenters. For each subject, the set of faces chosen for the PET experiment was composed from those that the subject recognized. All other items were derived from the standardized set of stimuli used in our laboratory. In the PET scanning session, subjects performed each task twice, in random order, from 5 seconds until 65 seconds after injection of $[^{15}O]$ H_2O. Subjects named approximately 85 percent of stimuli in each category, and task performance success was not significantly different among the naming tasks (nonresponse rates for naming persons, animals, and tools were 17.0%, 14.8%, and 11.7%).

Procedure   PET data were acquired with a General Electric 4096 Plus whole body tomograph, yielding 15 transaxial slices with a nominal interslice interval of 6.5 mm and an intrinsic resolution of 6.5 mm in all axes using the $[^{15}O]H_2O$ bolus technique. The labelled water was continuously available in the scanner room. Approximately 50 mCi of $[^{15}O]H_2O$ was administered for each of eight injections. Regional cerebral blood flow was estimated according to the autoradiographic method (Herscovitch, Markham, and Raichle 1983; Hichwa, Ponto, and Watkins 1995).

All subjects also underwent MR procedure using the parameters mentioned in the section on lesion studies. The MR images were used to reconstruct the brain in 3-D. This data set was used to orient the PET slices parallel to the longitudinal axis of the temporal lobe, in an orientation parallel to the plane that would intercept both the lowest points in the temporal poles and the lowest points in the occipital lobes. It also allowed us to define our region of interest.

We restricted the search volume, in the first phase of the study, to the overall region hypothesized for the lesion study: IT and TP. The limits of the TP/IT region were defined as the middle, inferior, and fourth temporal gyri, bounded superolaterally by the superior temporal sulcus and inferomesially by the collateral sulcus. The posterior boundary of the temporal lobe conformed to the definition in the atlas of Ono, Kubik, and Abernathy (1990). Anteriorly, TP was separated from the superior temporal gyrus by a plane perpendicular to the longitudinal axis of the temporal lobe at the level of the anterior ascending ramus of the Sylvian fissure.

The search volume used in the PET analysis was defined as all stereotactic voxels that corresponded to TP/IT in the majority of the subjects, and was determined as follows: (1) Talairach space was constructed for each subject (Talairach and Tournoux 1988; Grabowski, Damasio, and Damasio 1995); (2) The TP/IT regions of interest of each subject were converted into binary volumes, Talairach-transformed, and summed. Pixels with values reflecting the overlap of at least five subjects comprised the search volume.

The collection of 3-D MR data enabled us to avoid several potential technical artifacts in this study. We eliminated extracerebral pixels from the data set and we included in the search volume only those pixels which corresponded to TP/IT in the majority of the subjects. We are confident that all of the regions of activation which we detected were either exclusively in TP/IT or, if they spilled beyond its boundary, were centered within it.

Data Analysis   MR and PET data were coregistered using PET-Brainvox (H. Damasio *et al.* 1993; Grabowski, Damasio, and Damasio 1995) and Automated Image Registration (AIR) (Woods, Mazziotta, and Cherry 1993). PET data were subjected to Talairach transformation and then smoothed with an isotropic 16 mm Gaussian kernel. The final calculated image resolution was $19 \times 19 \times 18$ mm. PET data were analyzed with a pixelwise linear model which estimated coefficients for global flow (covariable) and task and block/subject effects (classification variables) (Friston *et al.* 1995; Grabowski *et al.* 1996). Global flow did not differ significantly across tasks. We compared adjusted mean activity in each of the three naming conditions to the control task using pixelwise *t*-statistics.

The common intracerebral stereotactic volume was 1106.3 cm³. The search volume (bilateral TP/IT) was 123.4 cm³. Using Worseley's theory of Gaussian fields (Worseley *et al.* 1992; Worseley 1994), the threshold *t* value (volumetric alpha 0.05, 58 degrees of freedom, 19 resels) was 3.82. We also analyzed the *t* fields using spatial extent rather than intensity thresholding.

Results

The results are very much concordant with those obtained with the lesion method. We do find the left temporal pole active during the retrieval of names for unique persons and two regions in the inferotemporal cortex active during the retrieval of words denoting animals and the retrieval of words denoting tools, thus supporting our initial hypothesis.

However, this is not all. If instead of limiting the search to the regions initially hypothesized we look at the whole brain volume, we also see differential activation in the left frontal operculum.

As for IT, a region of interest, the inferior and middle frontal gyri was identified in all subjects and used as the search volume. A pixelwise analysis showed that, in fact, there were three distinct foci of activation: one centered in the pars triangularis (Brodmann's area 45) present for all three categories; two other ones, one in prefrontal region present for word retrieval of unique persons; and another, in premotor cortex in the precentral sulcus, present for word retrieval of tools.

A further analysis of this premotor activation using a different approach, local standard space, was then performed. This method allows the analysis of PET data with respect to the distribution of flow values in each subject along an axis of a region identified for each subject (Grabowski *et al.* 1995). We delineated, for each subject, the area encompassing the gray matter of the precentral sulcus between the Sylvian fissure and the superior frontal sulcus. A

2-D Cartesian coordinate system was imposed on this region with a Y-axis extending between the upper and lower limits mentioned above and an X-axis perpendicular to the Y-axis. For this type of analysis no spatial filtering of the data is performed, and analysis is done using the pixelwise linear model employed for Talairach space (see details in Grabowski, Damasio, and Damasio 1998). This analysis highlighted the differential distribution of flow values in the precentral sulcus region for the word retrieval of tools. It is also interesting to note that the Talairach coordinates of this particular region of activation are very close to the region where activation is found during verb generation tasks (Petersen *et al.* 1998; Grabowski *et al.* 1996). In other words, activations related to words for actions and for entities with characteristic actions, as in the case of manipulable tools, are clustered about the same region.

Moreover, we also have described subjects with lesions in this area who show deficits in the retrieval of words denoting actions (A. R. Damasio and Tranel 1993).

Conclusion

From the data presented we can draw the following conclusions:

1. The neural support for normal language processes, even as assessed by one single aspect of language function, such as word retrieval, does not depend only on the two classic "language areas". A number of other areas are necessary to support normal language processing.

2. Given the same entity, damage to certain areas will impair naming but not preclude satisfactory concept retrieval; while damage to other areas will impair concept retrieval and thus naming, too. Given the same entity, it is possible to separate components of the system primarily dedicated to word retrieval or to concept retrieval.

3. The optimal retrieval of words for entities belonging to varied conceptual categories depends on anatomically segregated regions. This suggests that the systems which support word retrieval for varied categories are, at least in part, segregated.

4. The same applies to the retrieval of concepts themselves. Different regions are consistently associated with the retrieval of concepts for entities within certain categories.

5. The anatomical regions identified by lesions (e.g. *dysfunction* sites) are consistent with the anatomical regions identified in PET (e.g. *activation* sites).

8 What Thought Requires

Donald Davidson

The true fly is dipterous, but it has halteres which have evolved from posterior wings. It had been thought that the astonishing rapidity with which a fly maneuvers must be due to a direct neural connection between the eyes and the wings, but recent experiments at Berkeley suggest something more sophisticated (Chan, Prete, and Dickinson 1998). It was known before that the halteres, which beat antiphase to the wings, act as gyroscopes which stabilize flight on all three axes by feeding information directly to the wing muscles. Remove a fly's halteres and it crashes. What is new is that apparently the visual system is directly connected to the halteres, which then control the wing muscles. This fancy setup distinguishes between the aerodynamic forces and the Coriolis forces acting on the wings, permitting the fly to evade the flyswatter with marvelous ease.

The fly serves to remind us that an organism can discriminate aspects of its environment with superb accuracy, and make use of the resulting information in complex ways that help keep it alive, without anything we would, or should, call thought.

Leibniz, who believed animals are machines, was asked why he was so reluctant to kill a bothersome fly. A fly is just a machine, Leibniz replied, but what a wonderful machine! It certainly would not be right, he thought, to destroy a manmade machine of comparable complexity (Guhrauer 1846: ii. 364). Leibniz was perhaps more right than he realized: designing such a machine is surely beyond the dreams of even today's technology. Leibniz also saw a profound difference between the fly (or any non-human creature) and man: man thinks, the fly does not. The fly's reaction to visual input is far too rapid to involve thought. On the other hand, if a man managed to design such a machine, he might well be inclined to view his mechanical fly as calculating aerodynamic and Coriolis forces in order to maintain its stability during a double Immelman. But of course a machine that could do all that a fly can do, and no more, would not be calculating in the sense of giving conscious thought to the matter. So we need to ask what would turn calculation, in the sense in which a fly or a computer can calculate, into conscious thought?

We are just machines that are complex in ways flies are not, so the problem isn't one of transcending mere physical devices. I do not doubt that an artificer could, at least in principle, manufacture a thinking machine. The problem, for philosophy anyway, is what to aim for; what would show that the artificer had succeeded? I assume that you and I can tell, given enough time,.

and the right sort of environment, whether an object can think, and we can tell this without any clear idea of what is inside the skin. In this respect, Turing had the right idea, though his test was not conclusive for a variety of reasons. But what, more exactly, is it that we detect when we recognize an object as a thinking being?

Animals show by their behavior that they are making fine distinctions, and many of the things they discriminate we do too. They recognize individual people and other animals, distinguish between various sorts of animal, find their way back to places they have been before, and can learn all sorts of tricks. So it is important to reflect on why none of this shows they have propositional attitudes: beliefs, desires, doubts, intentions, and the rest. They see and hear and smell all sorts of things, but they do not perceive *that* anything is the case. Some animals can learn a great deal, but they do not learn *that* something is true.

Why doesn't the fact that a horse or a duck discriminates many of the things we do strongly suggest that they have the same concepts we do, or at least concepts much like ours? This is a suggestion many find persuasive, and it is apparently unavoidably seductive for most of us when we describe the activities of dumb beasts. But there is little reason to take the suggestion literally. Someone could easily teach me to recognize a planet in our solar system (smaller than the sun and moon, untwinkling) without my having any idea what a planet is. A horse can distinguish men from other animals, but if it has a concept of what it is distinguishing, that concept is nothing like ours. Our concept is complicated and rich: we would deny that someone had the concept of a man who did not know something about what distinguishes a man from a woman, who did not know that fathers are men, that every man has a father and a mother, and that normal adults have thoughts. Creatures with propositional attitudes and creatures without are alike in that both can be conditioned to respond differentially to many of the same properties, objects, or types of event. This misleads us into thinking similar processes are going on in the brains of both sorts of creature. But a creature with propositional attitudes is equipped to fit a new concept into a complex scheme in which concepts have logical and other relations to one another. Speechless creatures lack the conceptual framework which supports propositional attitudes.

I think this is enough to ensure that some degree of holism goes with having concepts. Many concepts are fairly directly connected, through causality, with the world, but they would not be the concepts they are without their connections with other concepts, and without any relations to other concepts, they would not be concepts. To say this is neither to suppose holism is so pervasive that no two people could, in any sense required for communication, have the same concept, nor is it to deny that the contents of some concepts are more directly attached to sensory moorings than others. We can appreciate why holism is not the disaster it has sometimes been portrayed as being if, instead of asking how the content of a concept or judgment looks from the inside, we ask instead how an observer can size up the contents of the thoughts of another creature. This is, again, the Turing approach. So here I want to say something about how I think it is possible for one creature with a full basic set of concepts to come to understand another, for I think this will throw light on the central question I raised, which is how we can tell when a creature has a genuine concept.

There is no distinction to be made between having concepts and having propositional attitudes. To have a concept is to class things under it. This is not just a matter of being natively disposed, or having learned, to react in some specific way to items that fall under a concept; it is to *judge* or *believe* that certain items fall under the concept. If we do not make this a condition on having a concept, we will have to treat simple tendencies to eat berries, or to seek warmth and avoid cold, as having the concepts of a berry, or of warm, or of cold. I assume we don't want to view earthworms and sunflowers as having concepts. This would be a terminological mistake, for it would be to lose track of the fundamental distinction between a mindless disposition to respond differentially to the members of a class of stimuli, and a disposition to respond to those items *as* members of that class.

Given the task of deciphering a language we do not know, we will perforce start with perception sentences, sentences which a speaker will assent to or dissent from given a stimulus we too can sense. As Quine put it, "Linguistically, and hence conceptually, the things in sharpest focus are the things that are public enough to be talked of publicly, common and conspicuous enough to be talked of often, and near enough to sense to be quickly identified and learned by name" (Quine 1960: 1). Our first guess as to what is meant by a perception sentence will be a shot in the dark, but, given how much alike people are, getting it right is like hitting a barn door; the most casual guess is often correct. The simplicity of this mode of entry into an alien language should not leave us thinking that a concept so identified is *defined* by its external causes, without the aid of theory or a supporting nexus of further concepts. A concept is defined by its typical cause only within the framework of a system of concepts that allows us to respond to certain stimuli as tables, friends, horses, and flies. A concept is defined *for us* by its typical causes, given that we are already in the world of language and conceptualization. But patterns of stimulation do not, in themselves, delineate the content of any sentence or concept. Only a very modest degree of holism is enough to lead to the conclusion that no simple story about the causal relations between mental states and the world can account for intentionality, much less specify the intentional contents of thoughts or utterances.

Concepts, and the sentences and thoughts that employ them, are in part individuated by their causal relations to the world and in part by their relations to each other. Thoughts, because they have propositional content, are unlike everything else in the world except for utterances in having logical relations to each other. There is only one way for an interpreter to spot these relations, and that is by noting patterns among the utterances to which a speaker awards credence. Thus the interpreter will note that a speaker who assents to "It rained in Spain and we all got wet" will also assent to "It rained in Spain" and "We all got wet"; that a speaker who assents to "John is taller than Sam" will also assent to "Sam is shorter than John"; and so on. These examples illustrate the routes to different discoveries. The pattern of the first example holds no matter what sentences are substituted for "It rained in Spain" and "We all got wet", and so leads to the identification of the truth functional connective for conjunction, one of the logical constants. The second pattern holds no matter what names are substituted for "John" and "Sam", and so leads to the recognition of a logical relation between the two two-place predicates, "taller than" and "shorter than". The former discovery is far more important, since it uncovers one of the most basic sources of the creativity of language, a recursive rule. One can easily see how the other logical constants, at least those involved in the first-order predicate calculus, can be identified.

A more subtle problem is that of discerning relations of evidential support among sentences, as viewed by the speaker, of course. These relations can be uncovered, but only by invoking a version of decision theory which, by finding the subjective probabilities of sentences, allows the computation of conditional probabilities. Degrees of evidential support, while more variable from speaker to speaker than matters of logic and logical form, are essential to the identification of theoretical terms less directly keyed to perception than perceptual sentences, for they provide the ties that give substance, along with the structure provided by theory, to theoretical concepts.

I have been pursuing the twin questions of the relations between thought and language and the world on the one hand, and the sort of structure thought and language require on the other, in order to evaluate claims that one analysis or another of thought or language is satisfactory, or to decide what criteria to employ in judging whether a creature or device is thinking. How much structure should we demand? Here the fact that the structure of language mirrors the structure of propositional thought is a help. Possession of a concept already implies a degree of creativity, since the point of a concept is that it is applicable to any item in an indefinitely large class. The fixed singular terms of a language are presumably finite in number, but demonstrative devices, whether combined with sortal predicates or not, provide the means for picking out an unlimited number of items. The truth-functional connectives, with their iterative powers, supply a further form of creativity. But is creativity enough? There is a good reason to think not. Consider a language consisting only of names, predicates, and the pure sentential connectives. Such a language has a finite vocabulary, but a potential infinity of sentences, so it is creative. But it is easy to give the semantics of such a language without introducing a concept of reference, and so without matching up either names or predicates with objects. The explanation is simple: given a finite vocabulary of names and predicates, the truth conditions of each of the finite number of sentences formed without the aid of connectives can be stated without considering the roles of the parts of sentences; the rest is truth tables. There is no reason to credit a creature with so simple a language with an ontology.

Should we nevertheless say a creature with these conceptual resources and no others has what we would call a language, or thoughts? It seems to me not. A creature without the concept of an object, however good it is at discriminating what we call objects, is a creature without even the rudiments of the framework of thought. What calls for ontology is the apparatus of pronouns and cross reference in natural languages, what we represent by the symbolism of quantifiers and variables in elementary logic. These devices provide the resources for constructing complex predicates, and at this point semantics must map names and predicates on to objects.

If I am right that language and thought require the structure provided by a logic of quantification, what further conceptual resources is it reasonable to consider basic? I have no definite list in mind, but if the ontology includes macroscopic physical objects, including animals, as I think it must, then there must be sortal concepts for classifying the items in the ontology. There must be concepts for marking spatial and temporal position. There must be concepts for some of the evident properties of objects, and for expressing the various changes and activities of objects. If such changes and activities can be characterized in turn, then the ontology must also include events, and among the concepts must be that of the relation between cause and effect. I am inclined to make some major additions to this list, as I shall indicate in a moment, but this is enough to suggest that the domain in which thought can occur is fairly complex. It is the domain each of us inhabits, but one we have good reason to suppose is inhabited by no other animal on earth, or machine.

Much of what I have said about the complexity and specificity of thought may be thought to be appropriate in connection with human thought, but applicable only in that context. In other words, I am revealing a provincial attitude toward intensionality. Well, perhaps. There is a further consideration, however, that may reinforce the anthropomorphic perspective. The important question, after all, isn't whether some animals have a simpler or degraded set of concepts; it is the question whether they have concepts at all. There is a clear difference between being disposed to react in different ways to Vs and Ws, as octopodes can be trained to do, and having concepts, however vague and poor, of those letters. To have a concept is to classify items as instantiating the concept or not, to judge, however implicitly, that here is a V and there is a W. The difference, as is well known, lies in the idea of error. We can say, if we like, that the octopus has erred when it reacts to the V as it was trained to react to the W. That was not what *we* had in mind, and its action may deprive the octopus of a tasty reward. But on what grounds can we claim that the octopus did not grasp our concept? As Wittgenstein says, whatever the octopus does is in accord with some rule, that is, some concept or other, which is a way of saying there is no reason to suppose it has *any* concept. What the octopus did, when it chose a V when we had trained it to chose the W, was not in accord with our idea of what resembles prior stimuli. But the judgment of resemblance is ours, not that of the octopus. So far as I can see, no account of error that depends on the classifications we find most natural, and counts what deviates from such as error, will get at the essence of error, which is that the creature itself must be able to recognize error. A creature that has a concept knows that the concept applies to things independently of what it believes. A creature that cannot entertain the thought that it may be wrong has no concepts, no thoughts. To this extent, the possibility of thought depends on the idea of objective truth, of there being a way things are which is not up to us. I do not see how any causal story about the sequence of stimuli reaching an isolated creature can account for this feature of conceptualization or intensionality, provided the story is told in the vocabulary of the natural sciences.

There is, I believe, a direction in which to look for a solution, and that direction has been pointed out by Wittgenstein. What is needed is something that can provide a standard against which an individual can check his or her reactions, and only other individuals can do this. To take the simplest case, consider two individuals jointly interacting with some aspect of the world. When the pair spot a lion, each hides behind a tree. If the individuals are in sight of one another, each also sees the other hide. Each is therefore in a position to correlate what he sees (the lion) with the other's reaction. After a time, a consequence is that if one individual sees a lion when the other does not, the one who does not see the lion is apt to treat the first's reaction as a conditioned stimulus, and also hide. Now consider a situation in which each sees the same lion, but one of the individuals, because the light is poor, or a tree partially obscures the lion, reacts as he normally reacts to a gazelle. This turns out to be a mistake. This little skit cannot, in itself, explain conceptualization or grasp of the idea of error on the part of either observer. It does no more than indicate the sort of conditions in which the idea of error could arise. Thus it suggests necessary (though certainly not sufficient) conditions for conceptualization.

Tyler Burge (1986) has argued that the content of a perceptual belief is the usual or normal cause of that belief. Thus the cause of the belief that a lion is now present is past correlations of lions with stimuli similar to the present stimulus. The difficulty with this proposal is that equally good answers would be that beliefs about lions are caused by the appropriate stimulation of the sense organs, or by the photons streaming from lion to eye, in which case the beliefs would be about stimulations or photons. There are endless such causal explanations, and each would dictate a different content for the same perceptual belief. It is natural to reply, and Burge does reply, that we have no idea how to characterize the various patterns of stimulated optic nerves that would be caused by a lion, aside from the way I just did it, by appealing to the role of lions. The force of this reply depends on the fact that we happen to have a single lion-concept, but no single concept for patterns of lion-caused firings of neurons. But nature with its causal doings is indifferent to our supply of concepts. When it is conceptualization that is to be explained, it begs the question to project our classifications on to nature.

Burge's suggestion fails two tests: it fails to pick out the relevant cause, and so gives no account of the content of perceptual sentences, and it fails to explain error. Adding a second person helps on both counts. It narrows down the relevant cause to the nearest cause common to two agents who are triangulating the cause by jointly observing an object and each other's reactions. The two observers don't share neural firings or incoming photons; the nearest thing they share is the object prompting both to react in ways the other can note. This is not enough to define the concept, as I said before, since to have the concept of a lion or of anything else is to have a network of interrelated concepts of the right sorts. But given such a network, triangulation will pick out the right content for perceptual beliefs. Triangulation also creates the space needed for error, not by deciding what an error is in any particular case, but by making objectivity dependent on intersubjectivity.

It is clear that for triangulation to work the creatures involved must be very much alike. They must class together the same distal stimuli, among them each others' reactions to those stimuli. In the end, it is just this double sharing of propensities that gives meaning to the idea of classing things together. We say: that creature puts lions together into a class. How do we tell? The creature reacts in relevantly similar ways to lions. What makes the responses similar? Our concepts do; we have the concepts that define these classes. It takes another creature enough like the first to see and say this. The sharing of many discriminatory abilities explains why a considerable degree of holism is no obstacle to communication. This is also why Turing had the right idea about how to tell if a device (or animal) is thinking.

Here we have a reason why the third person approach to language is not a mere philosophical exercise. The point of the study of radical interpretation is to grasp how it is possible for one person to come to understand the speech and thoughts of another, for this ability is basic to our sense of a world independent of ourselves, and hence to the possibility of thought itself. The third person approach is yours and mine.

Triangulation depends not only on a minimum of two creatures, but equally on shared external promptings. For this reason, among others, I think Kripke's account (1982) of what he takes to be Wittgenstein's "skeptical solution" to the puzzle of rule-following is inadequate to serve as the whole story about conceptualization. The problem is just the one we have been discussing: how to account for failure to apply a concept correctly, given that what one person might count as an error may just be another person applying a different concept. Kripke's suggestion is that if a learner fails to apply a concept as his teacher thinks correct, the learner has made a mistake. Unfortunately this does not distinguish between failure to apply a concept incorrectly and applying a different concept correctly—the very distinction in need of explication. But Kripke's examples have another, related, flaw: they concern mathematical examples, and so lack the shared stimulus to provide the possibility of a shared content.

Ostensive learning, whether undertaken by a radical interpreter as a first step into a second language, or undergone by someone acquiring a first language, is an example of triangulation. The radical interpreter has, of course, the idea of possible error, and so do his informants, and he can assume he and they share most basic concepts. Thus a first guess is apt to be right, though there can be no assurance of this in particular cases. Someone being initiated into the wonders of language and serious thought is also being initiated into the distinction between belief and knowledge, appearance and reality—in other words, the idea of error. Like triangulation, ostensive learning runs the risk of leaving unclear not only how the next step should go, but also what constitutes a wrong application of a concept at earlier steps. But neither the novice nor the sophisticated radical interpreter is in a position to question a teacher's or informant's early applications of a concept or word new to the learner. Teacher or informant may not be applying her own concepts correctly, but learner and interpreter must accept wrong steps as right until later in the game, since for them a concept is being given content. Erroneous ostensions on the teacher's part just lead the learner to learn a different concept from the one the teacher wished to introduce, and so will promulgate misunderstandings.

How will the learner or interpreter discover when he is applying a different concept than the one his teacher or informant had in mind, and when one of them is misapplying the same concept? There are various possible answers. Some will appeal simply to the power of consensus; but this cannot be conclusive. Of course, consensus of *use*, where use is assumed to reflect what the teacher or society *means*, is just what the learner or radical interpreter needs to recognize, but consensus of *application* does not distinguish the two varieties of error. As far as I can see, nothing in the observable behavior of teacher or learner with respect to an isolated sentence can sort this out. The distinction depends on relations among uttered sentences. The relation of evidential support among sentences provides powerful clues. When the learner says "That's a cow" when faced by a bull, is she erroneously applying the concept cow, or correctly applying a concept that covers both cows and bulls? If she also learns what may be the truth-conditions of "That's an udder", one can test whether she assents to "That's an udder" when presented with a cow but usually not when presented with a bull.

But the real test, in my opinion, is learning to *explain* errors. It is when one has learned to say or to think, "That looks green," "That man seems small," "I thought it was an oasis" when one has said or thought that something blue was green, or that the large man in the distance was small, or that what looked like an oasis was a mirage, that one has truly mastered the distinction between appearance and reality, between believing truly and believing falsely. It is also at this point, of course, that the distinction becomes clear between falsely thinking a bull is a cow, and simply applying the word "cow" to both.

Cognitive science aims, among other things, to deal with thought and thinking. Up to this point I have been chiefly concerned to speculate about the conditions thought must satisfy, about what constitutes the subject matter, or part of the subject matter, of cognitive science. But cognitive science also aims to be a science, or at least to be scientific. Is thought, as I have described it, amenable to scientific study?

One reasonable demand on a scientific theory is that it should be possible to define a structure in such a way that instances of that structure can be identified empirically. This requires laws, or generalizations, which predict what will be observed given observed input. Some of the most impressive early work in psychology satisfied this condition. It was found, for example, that if a subject—just about any subject who was not deaf—was repeatedly asked to adjust a variable tone so that it sounded half way between two fixed tones, the subject made decisions that consistently defined an interval scale, in other words, a scale formally like the ordinary scales for temperature. This is not psychophysical measurement, since it does not relate a physical magnitude with subjective judgments; the tonal scale simply relates subjective judgments with other subjective judgments. Patterns of such judgments instantiate the laws specified by the axioms which define an interval scale.

Bayesian decision theory, in the form which Frank Ramsey gave it (1990), is more subtle, but it is similar in that it relates judgments to judgments. Ramsey showed how, given only choices between wagers, it was possible to construct two scales, one for degrees of belief (sometimes called subjective probabilities), and one for comparative degrees of perceived value. It is a question, of course, whether anyone's choices satisfy the conditions necessary for constructing such scales. As is well known, this is a very tricky question because of the mixture of normative and descriptive elements that enter into an attempt to give empirical application to the theory. Actual tests of Bayesian decision theory seldom show perfect consistency with the conditions; on the other hand, neither do they show many absurd deviations from them, and there are often persuasive arguments to explain the deviations. What one can say is that "given the right conditions", *ceteris paribus*, the laws of decision theory do describe how people make real choices. The fact is that we all depend on this. People will seldom risk their lives for a small reward, will pay quite a lot for a good chance at a large prize, and so forth. These are laws of human behavior, difficult as they sometimes are to apply.

The laws of logic are, in the same way, laws of thought—always, of course, given the right conditions, and so forth. Tarski-type truth definitions, modified to fit natural languages, describe the basic semantic structure that informs the human language ability. We do not know how to fit all the idioms of natural languages into the format Tarski provided, but a very impressive core can be handled. These three structures, of logic, decision theory, and formal semantics, have the characteristics of serious theories in science: they can be precisely, that is, axiomatically, stated, and, given empirical interpretation and input, they entail endless testable results. Furthermore, logic, semantics, and decision theory can be combined into a single unified theory of thought, decision, and language, as I have shown (D. Davidson 1990). This is to be expected. Decision theory extracts from simple choices subjective scales for probabilities, that is, degrees to which sentences are held to be true, and for values or the extent to which various states of affairs are held to be desirable. Radical interpretation, as I briefly described it above, extracts truth-conditions, that is, meanings, and belief from simple expressions of assent and dissent. Formal semantics has logic built in, so to speak, and so does decision theory in the version of Richard Jeffrey (1965). Uniting the theories depends on finding an appropriate empirical concept, and one such concept is the relation between an agent, the time and circumstances of utterance, and two sentences, one of which the agent would rather have true than the other. The protocols for testing such a theory are like the protocols in the testing of decision theory except that the choices which express preferences are treated as awaiting interpretation rather than as already interpreted. Given the richness of the structure of the unified theory, it is possible to derive the usual scales for subjective probability and desire, applied to sentences, and from these to determine the meanings of the sentences.

There is a widespread feeling among philosophers—a feeling with which I have a good deal of sympathy—that we will not really understand the intensional attitudes, conceptualization, or language, until we can give a purely extensional, physicalistic, account of them. Unless we can in this sense *reduce* the intensional to the extensional, the mental to the physical, so the theme runs, we will not see how psychology can be made a seriously scientific subject. Jerry Fodor argues that if intensional and semantic predicates "form a closed circle", that is, can't be reduced to physical predicates, this "appears to preclude a physicalistic ontology for psychology since if psychological states were physical then there would surely be physicalistically specifiable sufficient conditions for their instantiation. But it's arguable that if the ontology of psychology is not physicalistic, then there is no such science" (Fodor 1990: 51). Fodor seems to indicate in a footnote that he is aware that psychological states and events may be physically describable one by one even though mentalistic predicates are neither definitionally nor nomologically reducible to the vocabularies of the physical sciences. But if this is the case, as I have argued, then the issue is not ontological; the question just concerns vocabularies. Whether or not the ontology of psychology is physicalistic, my guess is that Fodor believes there can't be a science of psychology if its subject matter can't be reduced, either definitionally or nomologically, to that of the physical sciences.

I think there is a lot to this. Since psychology wants to explain perception, for example, it wants to explain how certain events physically described cause beliefs through the agency of the senses. Any laws concerning such interactions would, it seems, amount to partial nomological reductions of the mental vocabulary to the physical. (Psychophysical measurement has produced plenty of laws, but these typically deal with the relations between physical quantities and sensations, not thoughts.) Any really complete scientific psychology would have at many points to relate the mental and the physical, by which I mean events and states described both in psychological and in physicalistic terms. A lot depends, of course, on how strict one wants the laws of such a science to be. With an ample sprinkling of "other things being equal" and "under normal conditions" clauses, we constantly utilize "laws" that relate the mental and the physcial in everyday life. But here nothing like the laws of physics is in the cards.

I didn't finish discussing the unified theory of thought and language which I mentioned above. How much like a serious science could it be? Formally it's as clear and precise as any science. The difficulties lie in the application, the empirical interpretation. One trouble springs from its holism, though not quite in the way one would expect. All serious science is holistic. Whenever we assign a number to a physical magnitude we assume the correctness of the conditions which must hold to justify the form of measurement involved. In the ordinary measurement of length, for example, we assume that the relation of *longer than* is transitive. This assumption has no empirical content until we give an interpretation to this relation and, once we do this, we are assuming that the operation we have specified for determining that one object is longer than another holds for all objects under study. If the law of transitivity fails in a single case, the entire theory of measurement of length is false, and we are not justified talking of physical lengths. Once one considers the further conditions imposed by the theory, one appreciates the thoroughly holistic character of almost any physical theory. What makes the empirical application of decision theory or formal semantics tricky is that the norms of rationality apply to the subject matter. In deciding what a subject wants or thinks or means, we need to see their mental workings as more or less coherent if we are to assign contents to them at all, given that the contents are partly defined by their relations to one another. As in any science, we must be able to describe the evidence in terms the relevant theory accepts. The trouble with the study of thought is that the standards of rationality, outside of decision theory and logic at least, are not agreed upon. We cannot compare our standards with those of others without employing the very standards in question. This is a problem that does not arise when the subject matter is not psychological.

In one respect, the unified theory of thought and meaning which I described is a little better off than one might think. The important primitive term in that theory is the one expressing the attitude of preferring one sentence true rather than another. This is certainly a psychological concept, and a pretty complicated one. So there is no chance that the theory can be specified in physical terms. On the other hand, the theory is entirely stated in extensional terms. The relation of preferring true is a three-place relation between an agent and two sentences, and it holds no matter how those three entities are described. Of course, propositional attitudes are involved; they just aren't expressed, in the theory, in a way that individuates attitudes generally, and in a way that would make the theory circular. In testing the theory, one would have to devise a way of telling when an agent preferred one sentence true rather than another. This is not such a bad deal, for if the operation one hits on at first is wrong, the theory will yield nothing intelligible. The richness of the structure of thought and meaning will necessarily tease out a workable interpretation. This is the attitude we take to physical measurement, and, in ordinary life, the attitude we actually take to the understanding of others.

9 Things about Things

Daniel C. Dennett

Perhaps we can all agree that in order for intelligent activity to be produced by embodied nervous systems, those nervous systems have to have things in them that are about other things in the following minimal sense: there is *information* about these other things not just present but *usable* by the nervous system in its modulation of behavior. (There is information about the climatic history of a tree in its growth rings—the information is present, but not usable by the tree.) The disagreements set in when we start trying to characterize what these things-about-things are—are they "just" competences or dispositions embodied somehow (e.g. in connectionist networks) in the brain, or are they more properly mental representations, such as sentences in a language of thought, images, icons, maps, or other data structures? And if they are "symbols", how are they "grounded"? What, more specifically, is the analysis of the *aboutness* that these things must have? Is it genuine intentionality or mere *as if* intentionality? These oft-debated questions are, I think, the wrong questions to be concentrating on at this time, even if, "in the end", they make sense and deserve answers. These questions have thrived in the distorting context provided by two ubiquitous idealizing assumptions that we should try setting aside: an assumption about how to *capture content* and an assumption about how to isolate the vehicles of content from the "outside" world.

A Thing about Redheads

The first is the assumption that any such aboutness can be (and perhaps must be) captured in terms of *propositions*, or *intensions*—sometimes called *concepts*. What would an alternative claim be? Consider an old example of mine:

Suppose, for instance, that Pat says that Mike "has a thing about redheads." What Pat means, roughly, is that Mike has a stereotype of a redhead which is rather derogatory and which influences Mike's expectations about and interactions with redheads. It's not just that he's prejudiced against redheads, but that he has a rather idiosyncratic and *particular* thing about redheads. And Pat might be right—more right than he knew! It could turn out that Mike does have a thing, a bit of cognitive machinery, that is *about redheads* in the sense that it systematically comes into play whenever the topic is redheads or a redhead, and that adjusts various parameters of the cognitive machinery, making flattering hypotheses about redheads less likely to be entertained, or confirmed, making relatively aggressive behavior *vis-à-vis* redheads closer to

I want to thank Chris Westbury and Rick Griffin for comments on an earlier draft.

implementation than otherwise it would be, and so forth. Such a *thing about redheads* could be very complex in its operation or quite simple, and in either case its role could elude characterization in the format:

Mike believes that: $(x)(x$ is a redhead $\supset \ldots)$

no matter how deviously we piled on the exclusion clauses, qualifiers, probability operations, and other explicit adjusters of content. The contribution of Mike's thing about redheads could be perfectly determinate and also undeniably contentful and yet no linguification of it could be more than a mnemonic label for its role. In such a case we could say, as there is often reason to do, that various beliefs are *implicit* in the system. (Dennett 1987:148)

But if we do insist on recasting our description of the content in terms of implicit beliefs, this actually masks the functional structure of the things that are doing the work, and hence invites us to ask the wrong questions about how they work. Suppose we could "capture the content" of such a component by perfecting the expression of some sentence-implicitly-endorsed (and whether or not this might be "possible in principle", it is typically not remotely feasible). Still, our imagined triumph would not get us one step closer to understanding how the component accomplished this. After all, our model for such an activity is the interpretation of data structures in computer programs, and the effect of such user-friendly interpretations ("this is how you tell the computer to treat what follows as a comment, not an instruction to be obeyed") is that they direct the user/interpreter's attention *away from* the grubby details of performance by providing a somewhat distorted (and hyped up) sense of what the computer "understands". Computer programmers know enough not to devote labor to rendering the intentional interpretations of their products "precise" because they appreciate that these are mnemonic labels, not *specifications* of content that can be used the way a chemist uses formulae to describe molecules. By missing this trick, philosophers have created fantasy worlds of propositional activities marshaled to accomplish reference, recognition, expectation-generation, and so forth. What is somewhat odd is that these same philosophers have also largely ignored the areas of Artificial Intelligence that actually do take such content specifications seriously: the GOFAI worlds of expert systems, inference engines, and the techniques of resolution theorem-proving and the like. Presumably they can see at a glance that whatever these researchers are doing, their products are not remotely likely to serve as realistic models of cognitive processes in living minds.

A thing-about-redheads is not an axiomatized redhead-theory grafted into a large data base. We do not yet know how much can be done by a host of things-about-things of this ilk because we have not yet studied them directly, except in very simple models—such as the insectoid subsumption architectures of Rodney Brook and his colleagues. One of the chief theoretical interests of Brooks's Cog project is that it is pushing these profoundly non-propositional models of contentful structures into territory that is recognizable as human psychology. Let's see how they work, how they interact, and how much work they can do before we take on the task of linguifying their competences as a set of propositions-believed.

Transducers, Effectors, and Media

The second ubiquitous assumption is that we can think of a nervous system as an information network tied to the realities of the body at various restricted places: transducer or *input* nodes and effector or *output* nodes. In a computer, there is a neat boundary between the "outside" world and the information channels. A computer can have internal transducers too, such as a temperature transducer that informs it when it is getting too hot, or a transducer that warns it of irregularities in its power supply, but these count as *input* devices since they extract information from the (internal) environment and put it into the common medium of information-processing. It would be theoretically tidy if we could identify the same segregation of information channels from "outside" events in a body with a nervous system, so that all interactions happened at identifiable transducers and effectors. The division of labor this permits is often very illuminating. In modern machines it is often possible to isolate the control system from the system that is controlled, so that control systems can be readily interchanged with no loss of function. The familiar remote controllers of electronic appliances are obvious examples, and so are electronic ignition systems (replacing the old mechanical linkages) and other computer-chip-based devices in automobiles. And *up to a point*, the same freedom from particular media is a feature of animal nervous systems, whose parts can be quite clearly segregated into the peripheral transducers and effectors, and the intervening transmission pathways, which are all in the common medium of impulse trains in the axons of neurons.

At millions of points, the control system has to interface with the bodily parts being controlled, as well as with the environmental events that must be detected for control to be well-informed. In order to detect light, you need something photosensitive, something that will respond swiftly and reliably to photons, amplifying their sub-atomic arrival into larger-scale events that can trigger still further events. In order to identify and disable an antigen, for instance, you need an antibody that has the right chemical composition. Nothing else will do the job. It would be theoretically neat if we could segregate these points of crucial contact with the physics and chemistry of bodies, thereby leaving the rest of the control system, the "information-processing proper", to be embodied in whatever medium you like. After all, the power of information theory (and automata theory) is that they are entirely neutral about the media in which the information is carried, processed, stored. You can make computer signals out of anything—electrons or photons or slips of paper being passed among thousands of people in ballrooms. The very same algorithm or program can be executed in these vastly different media, and achieve the very same effects, if hooked up at the edges to the right equipment.

As I say, it would be theoretically elegant if we could carry out (even if only in our imagination) a complete segregation. In theory, every information-processing system is tied at both ends, you might say, to transducers and effectors whose physical composition is forced by the jobs that have to be done by them, but in between, everything is accomplished by medium-neutral processes. In theory, we could declare that what a mind *is* is just the control system of a body, and if we then declared the transducers and effectors to be just outside the mind proper—to be part of the body, instead—we could crisply declare that a mind can in principle be made out of anything, anything at all that had the requisite speed and reliability of information-handling.

This important theoretical idea sometimes leads to serious confusions, however. The most seductive confusion is what I call *the myth of double transduction*: first the nervous system transduces light, sound, temperature, and so forth into neural signals (trains of impulses in nerve fibers) and second, in some special central place, it transduces these trains of impulses into some *other* medium, the medium of consciousness! This is, in effect, what Descartes thought, and he declared the pineal gland, right in the center of the brain, to be the locus of that second transduction. While nobody today takes Descartes's model of the second transduction seriously, the idea that such a second transduction must somewhere occur (however distributed in the brain's inscrutable corridors) is still a powerfully attractive, and power-fully distorting, subliminal idea. After all (one is tempted to argue) the neuronal impulse trains in the visual pathways for seeing something green, or red, are practically indistinguishable from the neuronal impulse trains in the auditory pathways for hearing the sound of a trumpet, or a voice. These are mere transmission events, it seems, that need to be "decoded" into their respective visual and auditory events, in much the way a television set transduces some of the electromagnetic radiation it receives into sounds and some into pictures. How could it *not* be the case that these silent, colorless events are transduced into the bright, noisy world of conscious phenomeno-logy? This rhetorical question invites us to endorse the myth of double transduction in one form or another, but we must decline the invitation. As is so often the case, the secret to breaking the spell of an ancient puzzle is to take a rhetorical question, like this one, and decide to answer it. How could it not be the case? That is what we must see.

What is the literal truth in the case of the control systems for ships, automobiles, oil refineries, and other complex human artifacts doesn't stand up so well when we try to apply it to animals, not because minds, unlike other control systems, have to be made of particular materials in order to generate that special aura or buzz or whatever, but because minds have to interface with historically pre-existing control systems. Minds evolved as new, faster control systems in creatures that were already lavishly equipped with highly distributed control systems (such as their hormonal systems), so their minds had to be built on top of, and in deep collaboration with, these earlier systems.[1]

This distribution of responsibility throughout the body, this interpenetration of old and new media, makes the imagined segregation more misleading than useful. But still one can appreciate its allure. It has been tempting to argue that the observed dependencies on particular chemicals, and particular physical structures, are just historical accidents, part of an evolutionary legacy that might have been otherwise. True cognitive science (it has been claimed) ought to ignore these historical particularities and analyze the fundamental *logical* structure of the information-processing operations executed, independent of the hardware.

The Walking Encyclopedia

This chain of reasoning led to the creation of a curious intellectual artifact, or family of artifacts, that I call the *Walking Encyclopedia*. In America, almost every schoolyard has one student picked out by his classmates as the Walking Encyclopedia—the scholarly little fellow who knows it all, who answers all the teacher's questions, who can be counted on to know the capital cities of all the countries of the world, the periodic table of chemical elements, the dates of all the kings of France, and the scores of all the World Cup matches played during the last decade. His head is packed full of facts, which he can call up at a moment's notice to amaze or annoy his companions. Although admired by some, the Walking Encyclopedia is sometimes seen to be curiously misusing the gifts he was born with. I want to take this bit of folkloric wisdom and put it to a slightly different use: to poke fun at a vision of how a mind works.

According to this vision, a person, a living human body, is composed of a collection of transducers and effectors intervening between a mind and the world. A mind, then, is the control system of a vessel called a body; the mind is material—this is not dualism, in spite of what some of its ideological foes have declared—but its material details may be safely ignored, except at the interfaces—the overcoat of transducers and effectors. Figure 9.1 is a picture of the Walking Encyclopedia.

In this picture—there are many variations—we see that just inboard of the transducers are the perceptual analysis boxes that accept their input, and yield their output to what Jerry Fodor has called the "central arena of belief-fixation" (Fodor 1983). Just inboard of the effectors are the action-directing systems, which get their input from the planning department(s), interacting with the encyclopedia proper, the storehouse of world knowledge, via the central arena of belief-fixation. This crucial part of the system, which we

---

[1] The previous six paragraphs are drawn, with some revisions and additions, from Dennett 1996.

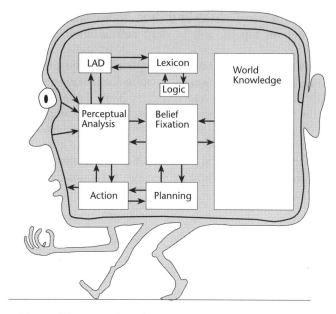

Figure 9.1   The Walking Encyclopedia.

might call the thinker, or perhaps the cognition chamber, updates, tends, searches, and—in general—exploits and manages the encyclopedia. Logic is the module that governs the thinker's activities, and Noam Chomsky's LAD, the Language Acquisition Device, with its Lexicon by its side, serves as a special purpose, somewhat insulated module for language entry and exit.

This is the generic vision of traditional cognitive science. For several decades, controversy has raged about the right way to draw the connecting boxes that compose the flow charts—the "boxology"—but little attention has been devoted to the overcoat. That is not to say that perception, for instance, was ignored—far from it. But people who were concerned with the *optics* of vision, or the *acoustics* of audition, or the *physics* of the muscles that control the eye, or the vocal tract, were seen as working on the *periphery* of cognitive science. Moreover, those who concerned themselves with the physics or chemistry of the activities of the central nervous system were seen to be analogous to electrical engineers (as contrasted with computer scientists).

We must not let this caricature get out of hand. Boxologists have typically been quite careful to insist that the interacting boxes in such flow diagrams are not supposed to be anatomically distinct subregions of the brain, separate organs or tissues "dedicated" (as one says in computer science) to the tasks inscribed in the boxes, but rather a sort of logical decomposition of the task into its fundamental components, which could then be executed by "virtual machines" whose neuroanatomical identification could be as inscrutable and gerrymandered as you like—just as the subroutines that compose a complex software application have no reserved home in the computer's hardware but get shunted around by the operating system as circumstances dictate.

The motivation for this vision is not hard to find. Most computer scientists don't really have to know anything much about electricity or silicon; they can concentrate on the higher, more abstract software levels of design. It takes both kinds of experts to build a computer: the concrete details of the hardware are best left to those who needn't concern themselves with algorithms or higher-level virtual machines, while voltages and heat-dispersion are ignorable by the software types. It would be elegant, as I said, if this division of labor worked in cognitive science as well as it does in computer science, and a version of it does have an important role to play in our efforts to reverse-engineer the human mind, but the fundamental insight has been misapplied. It is not that we have yet to find the right boxology; it is that this whole vision of what the proper functioning parts of the mind are is wrong. The right questions to ask are not:

How does the Thinker organize its search strategies?

or

Isn't the Lexicon really a part of the World Knowledge storehouse?

or

Do facts about the background have to pass through Belief Fixation in order to influence Planning, or is there a more direct route from World Knowledge?

These questions, and their kin, tend to ignore the all-important question of how subsystems could come into existence, and be maintained, in the highly idiosyncratic environment of a mammalian brain. They tend to presuppose that the brain is constructed of functional subsystems that are themselves designed to perform in just such an organization—an organization roughly like that of a firm, with a clear chain of command and reporting, and each sub-unit with a clear job description. We human beings do indeed often construct such artificial systems—virtual machines—in our own minds, but the way they come to be implemented in the brain is not how the brain itself came to be organized. The right questions to ask are about how *else* we might conceptualize the proper parts of a person.

Evolution embodies information in every part of every organism. A whale's baleen embodies information about the food it eats, and the liquid medium in which it finds its food. A bird's wing embodies information about the medium in which it does its work. A chameleon's skin, more dramatically, carries information about its current environment. An animal's viscera and hormonal systems embody a great deal of information about the world in which its ancestors have lived. This information doesn't have to be copied into the brain at all. It doesn't have to be "represented" in "data structures" in the nervous system. It can be exploited by the nervous system, however, which is designed to rely on, or exploit, the information in the hormonal systems just as it is designed to rely on, or exploit, the information embodied in the limbs and eyes. So there is wisdom, particularly about preferences, embodied in the rest of the body. By using the old bodily systems as a sort of sounding board, or reactive audience, or critic, the central nervous system can be guided—sometimes nudged, sometimes slammed—into wise policies. Put it to the vote of the body, in effect.[2]

Let us consider briefly just one aspect of how the body can contribute to the wise governance of a mind without its contribution being a data structure or a premise or a rule of grammar or a principle. When young children first encounter the world, their capacity for attending is problematic. They alternate between *attention-capture*—a state of being transfixed by some object of attention from which they are unable to deflect their attention until externally distracted by some more powerful and enticing signal—and *wandering* attention, attention skipping about too freely, too readily distracted. These contrasting modes are the effects of imbalances between two opponent processes, roughly captured under the headings of boredom and interest. These emotional states—or proto-emotional states, in the infant—play a heavy role in protecting the infant's cognitive systems from debilitating mismatches: when confronted with a problem of pattern-recognition that is just too difficult, given the current immature state of the system, boredom ensues, and the infant turns off, as we say. Or turns away, in random search of a task more commensurate with the current state of its epistemically hungry specialists. When a nice fit is discovered, interest or enthusiasm changes the balance, focusing attention and excluding, temporarily, the distractors.[3]

I suppose this sort of meta-control *might* in theory have been accomplished by some centralized executive monitor of system-match and system-mismatch, but in fact, it seems to be accomplished as a byproduct of more ancient, and more visceral, reactions to frustration. The moral of this story may not strike one as news until one reflects that nobody in traditional Artificial Intelligence or cognitive science would ever have suggested that it be important to build a capacity for boredom or enthusiasm into the control structure of an artificially intelligent agent.[4] We are now beginning to see, in

---

[2] The preceding two paragraphs are from Dennett 1996.

[3] Cynthia Ferrell, discussion at American Association for Artificial Intelligence Symposium on Embodied Cognition and Action, MIT, Nov. 1996.

[4] Consider a sort of problem that often arises for learning or problem-solving programs whose task can be characterized as "hill-climbing"—finding the global summit in a problem landscape pocked with lower, local maxima. Such systems have characteristic weaknesses in certain terrains, such as those with a high, steep, knife-edge "ridge" whose summit very gently slopes, say, east to the global summit. Whether to go east or west on the ridge is something that is "visible" to the myopic hill-climbing program only when it is perched right on the knife-edge; at every other location on the slopes, its direction of maximum slope (up the "fall line" as a skier would say), is roughly perpendicular to the desired direction, so such a system tends to go into an interminable round of overshooting, back and forth over the knife-edge, oblivious to the futility of its search. Trapped in such an environment, an otherwise powerful system becomes a liability. What one wants in such a situation, as

many different ways, how crippled a mind can be without a full complement of emotional susceptibilities.[5]

Things that Go Bump in the Head

But let me make the point in a deeper and more general context. We have just seen an example of an important type of phenomenon: the elevation of a byproduct of an existing process into a functioning component of a more sophisticated process. This is one of the royal roads of evolution.[6] The traditional engineering perspective on all the supposed subsystems of the mind—the modules and other boxes—has been to suppose that their intercommunications (when they talk to each other in one way or another) were not noisy. That is, although there was plenty of *designed* intercommunication, there was no *leakage*. The models never supposed that one box might have imposed on it the ruckus caused by a nearby activity in another box. By this tidy assumption, all such models forgo a tremendously important source of raw material for both learning and development. Or to put it in a slogan, such over-designed systems sweep away all opportunities for opportunism. What has heretofore been mere noise can be turned, on occasion, into signal. But if there is no noise—if the insulation between the chambers is too perfect—this can never happen. A good design principle to pursue, then, if you are trying to design a system that can improve itself indefinitely, is to equip all processes, at all levels, with "extraneous" byproducts. Let them make noises, cast shadows, or exude strange odors into the neighborhood; these broadcast effects willy-nilly carry information about the processes occurring inside. In nature, these broadcast byproducts come as a matter of course, and have to be positively shielded when they create too many problems; in the world of computer simulations, however, they are traditionally shunned—and would have to be willfully added as gratuitous excess effects, according to the common wisdom. But they provide the only sources of raw material for shaping into novel functionality.

It has been recognized for some time that randomness has its uses. For instance, sheer random noise can be useful in preventing the premature

---

Geoffrey Hinton has put it, is for the system to be capable of "noticing" that it has entered into such a repetitive loop, and resetting itself on a different course. Instead of building an eye to oversee this job, however, one can just let boredom ensue.

[5] Antonio Damasio's recent book *Descartes's Error* (A. R. Damasio 1994) is a particularly effective expression of the new-found appreciation of the role of emotions in the control of successful cognition. To be fair to poor old Descartes, however, we should note that even he saw—at least dimly—the importance of this union of body and mind: "By means of these feelings of pain, hunger, thirst, and so on, nature also teaches that I am present to my body not merely in the way a seaman is present to his ship, but that I am tightly joined and, so to speak, mingled together with it, so much so that I make up one single thing with it" (*Meditation* 6).

[6] In what follows I owe many insights to Lynn Stein's concept of "post-modular cognitive robotics" and to Eric Dedieu, "Contingency as a Motor for Robot Development", AAAI Symposium on Embodied Cognition and Action, MIT, Nov. 1996.

equilibrium of dynamical systems—it keeps them jiggling away, wandering instead of settling, until some better state can be found. This has become a common theme in discussions of these hot topics, but my point is somewhat different: My point is not that systems should make random noise—though this does have its uses, as just noted—but that systems should have squeaky joints, in effect, wherever there is a pattern of meaningful activity. The noise is not random from that system's point of view, but also not useful *to it*. A neighboring system may learn to "overhear" these activities, however, thereby exploiting it, turning into new functionality what had heretofore been noise.

This design desideratum highlights a shortcoming in most cognitive models: the absence of such noise. In a real hotel, the fact that the guests in one room can overhear the conversations in an adjacent room is a problem that requires substantial investment (in soundproofing) to overcome. In a virtual hotel, just the opposite is true: nobody will ever overhear anything from an "adjacent" phenomenon unless this is specifically provided for (a substantial investment). There is even a generic name for what must be provided: "collision detection". In the real world, collisions are automatically "detected"; when things impinge on each other they engage in multifarious interaction without any further ado; in virtual worlds, all such interactions have to be provided for, and most cognitive models thriftily leave these out—a false economy that is only now beginning to be recognized.

Efficient, effective evolution depends on having an abundant supply of raw material available to shape into new functional structures. This raw material has to come from somewhere, and either has paid for itself in earlier economies, or is a coincidental accompaniment of features that have paid for themselves up till then. Once one elevates this requirement to the importance it deserves, the task of designing (or reverse engineering) intelligent minds takes on a new dimension, a historical, opportunistic dimension. This is just one aspect of the importance of maintaining an evolutionary perspective on all questions about the design of a mind. After all, our minds had to evolve from simpler minds, and this brute historical fact puts some important constraints on what to look for in our own designs. Moreover, since learning in the individual must be, at bottom, an evolutionary process conducted on a different spatio-temporal scale, the same moral should be heeded by anybody trying to model the sorts of learning that go beyond the sort of parameter-tuning that is exhibited by self-training neural nets whose input and output nodes have significances assigned outside the model.

Conclusions

Cognitive science, like any other science, cannot proceed efficiently without large helpings of oversimplification, but the choices that have more or less defined the field are now beginning to look like false friends. I have tried to suggest some ways in which several of the traditional enabling assumptions of cognitive science—assumptions about which idealized (over-)simplifications will let us get on with the research—has sent us on wild goose chases. The "content capture" assumption has promoted the mis-motivated goal at *explicit expression* of content in lieu of the better goal of explicit models of functions that are only indirectly describable by content-labels. The "isolated vehicles" assumption has enabled the creation of many models, but these models have tended to be too "quiet", too clean for their own good. If we set these assumptions aside, we will have to take on others, for the world of cognition is too complicated to study in *all* its embodied particularity. There are good new candidates, however, for simple things about things now on offer. Let's give them a ride and see where we get.

10 On Referential Semantics and Cognitive Science

James T. Higginbotham

1. Introduction

Cognitive science is predicated on the view that mental states and processes (or many of them, and not only cognitive but also emotional states) are computational; that is, (*a*) that they are classified in part according to what they represent, and (*b*) that causal transactions between them have a purely formal basis. The states and processes of abstract machines, for instance in the sense of Eilenberg (1974), illustrate both features. What the machine is computing is determined by coding the inputs and decoding outputs, through codes that normally lend themselves to a coding of the internal configurations as well. But the states and configurations within the machine have what effects they do in virtue of their formal characteristics, indifferent to the coding, or what interpretation they may have. I am interested here in the role of referential semantics within cognitive science, so conceived.

I should add at the outset that I will be assuming also a view of cognitive science that admits, as targets for understanding and explanation, our ordinary, conscious mental states, and the cogency of common-sensical, rationalizing explanations of action, change of belief, and so forth. Part of the burden of my argument, indeed, will be that some conscious mental states involve a subject's conception of the meaning of an expression, and that rationalizing explanations of how one speaks, what one takes another to have said, and what meaning one conceives an expression to have will implicate referential semantics, under a conception of the subject that I advance below, as a subject for cognitive science. Skepticism about whether our most ordinary notions of belief, knowledge, and desire or interest will endure as part of the explananda of cognitive science, or rather will wither away as the subject progresses, will be put aside here.

Within semantic theory, as practiced in psychological, philosophical, and linguistic circles, there are inevitable differences of framework, modes of execution, and so forth. The outlines of referential semantics, however, are sufficiently constant as among these that distinctions will seldom matter for the purposes of the present discussion. In referential semantics, the central notion is that of the reference of an expression (for a person, at a time, and within an idiolect or style of speech for that person); the referential concepts comprise the reference of terms, and conditions on satisfaction and truth;

these conditions attach to the primitive expressions, or the lexicon, in virtue of their form; and they are tracked beyond the lexicon through complex linguistic structures, building up to conditions or constraints on the truth conditions of entire, closed sentences, or discourses.

Referential semantics, so conceived, can be completed as an account, or partial account, of communication, only as supplemented by, and perhaps appearing only as a part of, a fuller theory of how the exploitation of context in communicative exchange, the purposes and presuppositions of speakers and hearers, combine with the conditions given by language alone. In the view of Robert Stalnaker (1998), for instance, what language delivers is a content (a set of possible worlds), in a context (another set of possible worlds, conceived as the set of possibilities left open by the context); and a number of refinements on this view have been suggested. In all of these approaches, however, the notions of reference, satisfaction, and truth (relativized now to possible worlds, or to one or another substitute or additional parameter) are the fundamental ones, and these notions operate in the familiar way, associating parametric conditions on reference with the primitives, and projecting them through syntactic structures.

To say that reference is crucial is not to say that reference is all there is to meaning, or all there is to those aspects of meaning that are determined by linguistic form. Thus consider racial epithets. These are words that, having exactly the same conditions on reference as their more neutral counterparts, have very different conditions on their use. Euphemism, too, displays differences in meaning that are not differences in conditions on reference. In Britain, those who have retired at the end of their working lives and are living on accumulated pensions are referred to as "old age pensioners", and in official notices this reference is even acronymized, to OAP. Such usage, at least in public, is unthinkable in the United States, where reference to the old as such is felt to be demeaning, and where the euphemism "senior citizen" has long since been cleverly substituted. Well, but a senior citizen is an old person, and that is the end of that. There are several euphemisms for death, all of which are recognized as such by the people who use them, or know about the use that others make of them. Identity of conditions on reference is combined with differences in use in many similar cases: nicknames versus full names, and in general vernacular versus formal speech, and so on.

These features of use do not displace reference; on the contrary, they are understood only in the light of reference. The same may be said of several figures of speech, as noted in Goodman (1976). Hilary Putnam (1978: 99) once wrote that meaning should be viewed as a "coarse grid laid over use". What is right about this appealing image, if reference is the core of meaning, is the reverse of what it suggests. Meaning is not laid over use: use is a refinement of the grid laid down by reference. In any case, I shall consider here only such differences in meaning as are reflected in reference.

Given the broad conception of cognitive science sketched above, and the theory and practice of referential semantics as described, what role, if any, can referential semantics have within cognitive science? For all that I have said so far, it has no role whatever, since referential semantics is a theory of the structure and interpretation of actual or possible languages, and says nothing at all about the states of their speakers. To join the issue properly, we need a conception bridging between semantics and psychology. This conception, I assume (following Chomsky 1965, or extending his view to the semantic sphere), is that of knowledge of meaning as knowledge of conditions on reference. Thus I want to suppose that the speaker's relation to the statements of referential semantic theory is that of knowing them, and that grasp of the truth conditions of whole sentences is, for that person, a rational achievement made possible by the grasp of the conditions on reference of the primitive parts, and their mode of combination. The theory is a theory of semantic *competence*, to use Chomsky's word.

As you know, however, competence theories have been regarded with deep misgiving. In abstracting from the behavior and functional state of the subject except insofar as they provide evidence for the nature of competence, these theories leave the question of causal mechanism to one side, or rather replace it with inferential mechanisms whose causal basis, so far as the competence theory goes, is left blank. To be sure, we have intentional explanations of (some) linguistic behavior, constructed on the familiar cognition-and-conation model; but these fall short of what is required. In any case, the problem I wish to address is now set: what role can a theory of semantic competence, conceived as a theory of knowledge of reference, take within cognitive science?

In fact, if the premisses I have assumed be granted, the answer to the problem is straightforward. A person who knows a language is in a complex mental state that involves largely tacit knowledge of conditions on the reference of words, and principles that project these conditions through complex syntactic structures. This knowledge is deployed, if cognitive science has it right, through computation, acting on the representations of the speaker's tacit knowledge, and producing in consequence behavior and adjustments of states that go to exemplify our rationality in the use of language. Semantic theory (which must of course be articulated together with the theory of syntax) is concerned to identify the objects of knowledge, and it will be up to cognitive science, construed as the theory of the structure and causal links amongst our computational states, to say how it is realized. Part of the problem of syntax and semantics in cognitive science is then of the same shape as the problem of high-level vision. In the theory of vision we aim to go from the image received, via appropriate computations, to identification of the objects beheld. Similarly, in syntactic and semantic theory we need to go from the sequence of expressions to the syntactic structure that we apprehend, and the conditions on its interpretation that we know. The difference is that, in the case of language though not of vision, the objects that we know about are themselves known only through theoretical inquiry.

If I am right, then a defense of the conception just advanced of semantics within cognitive science will chiefly require a defense of its premises, that there is semantic competence, and that its backbone consists in knowledge of conditions on reference. My plan in what follows, therefore, will be to consider three (of the many) alternatives to these premises, endeavoring to show in each case that they are mistaken either in principle or in respect of the character of semantic theory. The first alternative, which I consider in a form I owe to Paul Horwich, suggests that semantics is concerned with the expression of concepts and propositions by words, and that the issue of compositionality, or projecting meaning from parts to wholes, is trivial, setting no constraints on the nature of a theory of meaning, and in particular not bringing in the notion of reference. The second, which can be identified as a commonplace of many writings in linguistics, is that linguistic theory is intrinsically part of cognitive science, because it is concerned with the nature of mental representations. If it were correct, then the path to integration of syntactic theory, if not semantics, would be very different from what I advocate above; but I shall argue that there is no reasonable way to construe it as consistent with the nature of linguistic competence, or indeed with the practice of linguistics itself. The third and last alternative, formulated and promoted most conspicuously by Jerry Fodor, holds that semantic theory is concerned not with knowledge, but rather with the interpretation of linguistic expressions as given by their translation into the language of thought. I will express doubts that this view can take in enough of what referential semantics has to consider; but I will observe in conclusion that the point of view that I take here is not only consistent with, but positively invites, a version of the representational theory of mind.

2.   Combinatorics

An assumption of referential semantics is that reference is the backbone of meaning, without which figures of speech, implicatures, etc., would be impossible. This assumption can be, and has been, questioned on the grounds that reference, satisfaction, and truth should be understood in a way that is "minimalist" in the sense of Horwich (1990) or "deflationist" in the sense of Field (1994) (so that they tell us nothing about meaning), and that meaning is constituted, not by conditions on reference, but by concepts expressed (where what concept a linguistic item expresses is derived in some way from its use, manifesting our practical abilities). A recent locus of this alternative is work by Paul Horwich (1997), entitled "The Composition of Meanings". I will set out what I take to be Horwich's position, and convey where, if my summary is correct, my disagreement would come. Whether I have him right or not, I think that the view that I will consider, and the rejoinder to it, deserve to be spelled out.

Meaning, however understood, must be compositional; that is, it must be seen to satisfy the general principle that the meaning of an expression is a function of the meanings of its parts and their mode of combination. The combinatorics of meaning is concerned with interpreting the notion of a "mode of combination", and with spelling out the cases that actually occur. Without loss of generality we can assume a version of compositionality that may be called *strictly local*; whether it is generally true or not, it certainly marks the default case.

Suppose that we are looking at a syntactic structure $Z = X - Y$, and that the meaning of whatever occupies X (the conditions on X's reference, together with whatever other information attaches to it) and of whatever occupies Y are known. The points X, Y, and Z will come with certain formal information (in the form, for instance, of syntactic features). Then the thesis of strict locality is that the meaning, or range of possible meanings, of Z is strictly determined by the formal information at X, Y, and Z, and the meanings of X and Y. The thesis implies, then, that further information about how the meanings of X and Y were arrived at is not relevant to the meaning of Z; and furthermore that the contribution of Z to any larger structure W within which it may occur is independent of the nature of W.

The configuration $Z=X-Y$, together with the formal features (identifying, say, X as a noun, Y as a verb, and Z as a sentence) constitutes a syntactic *schema*, to which a condition on interpretation is attached, by hypothesis. An *instance* of the schema is obtained by filling in X and Y with appropriate linguistic material. Let the schema be that of predication, illustrated by simple sentences such as 'Fido barks'. We know how to associate truth conditions with expressions falling under this schema, but the interest here is whether, by simply endowing the schema with a meaning of its own, we can express the combinatorics of this simple, and strictly local, case of compositionality without bringing truth conditions into it.

Suppose that (in virtue of whatever phenomena) we have fixed on the concepts expressed by the subject and predicate; say, that 'Fido' expresses FIDO (a certain individual concept) and 'barks' expresses BARKS. Then, the thought is, the meaning of the predication schema yields as the meaning of Z, the sequence 'Fido barks' with grammatical description as given by the syntax, the proposition FIDO BARKS. More generally:

if S is an instance 'a Vs' of the predication schema, the concept c_1 is the meaning of the instance of a, and c_2 is the meaning of the instance of V, then the meaning of S is (the proposition that) $c_1 c_2$.

What we have just said for predication would be applied across the board, throughout the various types of constructions: the meaning GREEN DOOR of 'green door' will be the result of applying the schema governing adjective +

noun to GREEN and DOOR; and so on. Compositionality, in the strictly local form, is therefore satisfied.

Notice that Horwich (or at least my version of his views) does not deny the thesis that the theory of meaning may be studied as the theory of a human competence. As such, it would have to explain how one may know the meaning of a complex expression given that one knows the meanings of its parts and their mode of combination; and in fact it purports to do just that, by means of the general condition above, for example, on combining concepts under the schema of predication.

However, as a theory of how you know what 'Fido barks' means, what that condition gives is unsatisfactory, since it does not determine what the meaning of the predication schema, or any of its instances, in fact is. This can be seen from the fact that a person who knew that 'Fido' means FIDO and 'barks' means BARKS, and apprehended the general statement above governing the predication schema would not in virtue of that come to know what the meaning of 'Fido barks' is, unless he already grasped that the latter is to be predicated of the former, which it was precisely the intention by means of that formula to convey. (The schema was *called* 'predication' in the formula, but that it *was* predication played no role in the formula.) But the meaning of 'Fido barks' must be specified as involving c_2 in the role of predicate of c_1, for we are juxtaposing names of concepts not by way of listing them, but by way of indicating certain propositions, in which c_2 is predicated of c_1.

In sum, one cannot specify the semantic effect that a mode of combination has on an ensemble of concepts simply by reproducing the linguistic specification of that mode. We know that FIDO BARKS is the proposition delivered by applying the predication schema to FIDO and BARKS (in that order); or alternatively, by applying BARKS to FIDO. But about the meaning of 'Fido barks' we know only that the schema yields FIDO BARKS as value, given FIDO and BARKS; and that does not tell us what proposition FIDO BARKS may be.

The above disquisition is intended as an echo of a swift remark of Donald Davidson (1967), rejecting the proposal that the meaning of the sentence 'Theaetetus flies' is the value of the meaning of 'flies' for the meaning of 'Theaetetus' as argument. Davidson writes: "The vacuity of this answer is obvious. We wanted to know what the meaning of 'Theaetetus flies' is; it is no good to be told that it is the meaning of 'Theaetetus flies'". (Davidson 1967: 20).

By way of contrast, compare how the interpretation of modes of combination, and the schema of predication in particular, proceed according to the account of meaning in terms of knowledge of reference. We assume that the native speaker knows

> 'Fido' refers to Fido
> 'barks' is true of x $\leftrightarrow$ x barks

and

> If S is an instance of '*a* Vs', the instance of '*a*' refers to y, and the instance
> of V is true of x $\leftrightarrow$ F(x), then S is true if and only if F(y).

Knowledge of all these (and the fact that 'Fido barks' is an instance of '*a* Vs',
where the instance of '*a*' is 'Fido' and that of V is 'barks') is sufficient to enable
the deduction concluding that 'Fido barks' is true if and only if Fido barks—
which was what the native speaker had to know to know the meaning of the
sentence.

I have spoken in terms of knowledge of reference, since I hold that (cases
where meaning is not simply referential aside, as discussed above) knowledge
of reference is sufficient for knowledge of meaning (or, rather, that the
relevant notion of meaning is exhausted thereby); but this was not essential.
Thus one could conceive an account whose principles were:

'Fido' means FIDO
'barks' means $^\wedge \lambda x(\text{BARKS}(x))$

and

> If S is an instance of '*a* Vs', the instance of '*a*' means *y*, and the instance of
> V means $^\wedge \lambda x(F(x))$ then S means F(y).

Since the context 'means . . .' is not extensional, more would have to be said in
order to enable the derivation of

'Fido barks' means that Fido barks

but it should be clear that there is a difference even between this formulation
and the trivial account that we are considering: the configuration that yielded
the proposition that Fido barks as the meaning of 'Fido barks' was not called
'the predication schema', but that it served *as* the predication schema follows
from the principle governing it (and of course makes our calling it the
predication schema redundant).

I conclude that attempts such as that I am attributing to Horwich have
obscured the point that a combinatorial semantic principle, if it is to confer
knowledge of the meaning of an expression on the basis of knowledge of the
meanings of its parts, cannot take the form envisaged. It is like an equation in
two unknowns; we know that the meaning of the predication schema is such
that it delivers FIDO BARKS given FIDO and BARKS; and we know that FIDO
BARKS is the value of the predication schema for those arguments; but from
this single principle we don't know what the predication schema means, or
what proposition FIDO BARKS may be.

3. Representationalism

Assuming now that knowledge of the conditions on the reference of words and syntactic structures is the core of knowledge of meaning, and that learning a language is a matter of learning such conditions, we can ask how such a knowledge-based conception might figure in cognitive science. One view, which has been prominent at least in the ideology of linguistics, is that tacit knowledge is to be integrated with cognitive science in virtue of its subject matter, namely mental representations. To see the rationale for this view, we may start with syntax, which has been said to be directly involved with mental representations. Not, indeed, that the statements of syntactic theory, fancy or plain, are statements *about* mental representations (prima facie at least, that would be absurd), but rather that they owe their truth to the internal workings of human beings, computationally described.

It may be thought that, if the statements of syntax are not statements about mental representations, then languages would not be empirical objects, and this consequence would vitiate the psychological interpretation of syntax. Nothing of the sort, so far as I can see, is true. Elsewhere (Higginbotham 1997) I have suggested that the objects of syntax are impure relational structures ("impure" as containing not only pure sets), so abstract objects; and this point of view itself is just a more contemporary elaboration of the view that they are sets of sequences of primitive marks. My speech, whatever it is, is one of these sets of structures, call it S*. But the identification of my syntax with S* is an empirical fact, calling for psychological explanation; and if this fact falls within cognitive science, then at least part of its basis lies in my computational mental structure.

The obvious point just made extends to semantics, as follows. That my word 'dog' refers to dogs is a fact about me which, if we had identified my language as the language with formal syntax S* and semantics S*', would follow from that very identification. The psychology comes, not in the formal derivation, but in the ascription to the speaker of the knowledge that supports it. Thus we are Platonistic with respect to languages, but linguistics itself is empirical.

But could syntactic and/or semantic theory be given another interpretation? We do not expect that our mental representations of syntactic structures bear no relation at all to the structures themselves; and the structures owe, not their existence, but anyway their interest to us, to the fact that they are structures that we somehow apprehend. Nevertheless, I see no way to reconcile humdrum linguistic statements, such as, "In NN's speech, the expression 'dog' is a Noun" with any such slogan as, "Syntactic theory is a theory of mental representations and processes." The word 'dog' (even my own particular word 'dog') is no mental representation of mine, although I presumably mentally represent *it* in some way; and the predicate 'Noun' is a predicate of expressions (for people, at times), and not of any mental representations. As George (1989) documents in detail, the conflation of grammar with what he calls psychogrammar, the nature of the mental representations of grammatical objects, is widespread, and can lead to misrepresentation of the contents of linguistic theory.

If failure to observe the distinction between grammar—what one's mental representations of linguistic objects represent—and psychogrammar—the representations that represent them—has been a source of obscurity in the psychological interpretation of linguistic theory, a still more problematic failure, in my opinion, has been the failure to distinguish between the deliverances of semantic theory and the ways in which they may be represented within us. I shall not document the failure here, but observe only that the widespread assumption that semantics is concerned with the nature of what are called "semantic representations" is a symptom of the problem. A semantic representation, whatever it is, is a representation of meaning, not a meaning; it is therefore an object of syntax, awaiting interpretation.

I do not mean to suggest that the idea that there are representations that have a special role in relation to semantics is intrinsically confused. On the contrary, in for example the theory of discourse representations as in the work of Hans Kamp, it is clear that the discourse representations are a syntax (one that incorporates various contextual features recovered on-line), and that accompanying the syntax there is a semantics, in fact a referential semantics in the sense of this essay. Likewise, it has been proposed that the linguistic level LF of Logical Form in the sense of Chomsky has a special status, in that interpretation, or certain aspects of it, is determined by the nature of representations at that level, which are remote from the representations immediately recovered when people speak. But the idea that a semantic representation may somehow go proxy for an interpretation is common enough, and this idea seems confused.

But what is wrong with it? One can propose, as I did above, that the notion of a semantic representation is either an oxymoron (if it were semantic, it would not be a representation, but rather a semantics, stating conditions on meaning), or else that the modifier 'semantic' is to be taken in an etiolated meaning, perhaps indicating only that the representation is the input to some features of the semantics. The problem lies deeper than that, however: for it may be said that the objects represented within the mind can only be syntactic, and that interpretation, conceived as relating representations to what they are about, cannot be found there. So representationalism would be right after all, not because the semantics of a language would be nothing more than its syntax, but because only the syntax is, properly speaking, an object of cognitive scientific inquiry.

Recall in this respect the picture of computation as involving input and output codes, and the inner workings of the computational device. The codes are part of the machine itself, and they determine what is being computed. The analogy to a computational theory of semantics is that the external world (and some interior states) are coded in some form as, say, the sight of a cat on a mat and a desire to tell the whereabouts of the cats in the vicinity; these, once coded up, are delivered to the language faculty, whose job it is to produce an appropriate representation of the sounds of speech; following this purely formal operation, the behavior of saying "Cat on Mat" follows, which is to be interpreted, or output-decoded, as saying that the cat is on the mat. The semantics then lies in the coding and decoding, not in language. But for a sophisticated version of this type of account, we may look to work especially by Jerry Fodor, to which I now turn.

4.   Translational Accounts of Semantics

In so general an inquiry as that of the constitution of the mental, the role of linguistic meaning might be held to be subordinate, and the role of detailed empirical investigation, of the sort typically conducted in linguistic theory, more recessive still. It may be questioned, for example, whether the link between the constituting phenomena and the constituted phenomenon of language is direct or indirect. Thus it is open to suggest that words mean what they do because of certain features of human activity involving words; that the primary activity behind the use of words is the conveyance of thoughts; and that once it has been explained what it is for us to have the thoughts we do, the further step to the meanings of words is straightforward, since the words merely serve as a medium for expressing the thoughts. This suggestion is most closely identified with the work of Paul Grice, and the general point of view has been characterized especially by Stephen Schiffer, who calls it "intention based semantics" (see Schiffer 1987; Grice 1989). But the general idea, that in the large project of integrating the mental into a comprehensive view of nature we should let language take a back seat in favor of thought, is one that has been supported from a variety of points of view. Jerry Fodor (1998: 9) has perhaps expressed it most provocatively in his suggestion that, properly understood, natural languages have no semantics. Semantics exists (and it is referential), but it is the semantics of mental representations. Human languages, in virtue of standing in some formal relation to these representations, inherit a semantics, but do not have it of their own.

Fodor's account, as he makes clear in the reference cited, combines one feature of intention-based semantics (namely, the thesis that knowing a language is a matter of associating linguistic expressions with thoughts) with the representational theory of mind, and in particular with the thesis that to have made such an association is to have translated the linguistic expressions into a medium, the language of thought, which in virtue of conspiracy between brain and world expresses the thought in question.

Of course, semantics must associate expressions (sentences) with thoughts: the question is how it is going to do this, and what the basis is for the association, and in particular whether ongoing linguistic practice exhausts the evidence, or whether not only the capacity for thought in general but also the capacity to think certain particular thoughts must be taken for granted from the beginning. However this may be, I want now to observe that there is nothing in the combination of the association of language with thought, together with the representational theory of mind, that precludes the view that semantic competence is a matter of knowledge of conditions on reference.

The reason is simple. Suppose, as on the no doubt idealized view of cognitive states that stems from the representational theory, that to stand in the relation of knowing so-and-so is to have, as it were in a box in one's head, a representation in the language of thought whose interpretation is so-and-so, where the label on the box expresses something like tacit belief, and the tacit belief is properly externally and internally grounded so that it counts as knowledge. This account of cognitive states is perfectly general, and so applies to states of knowledge of conditions on reference in particular. So nothing prevents, say, the basis of our capacity to associate the thought that Fido barks with the sentence 'Fido barks' from being the appearance in the tacit belief box, perhaps under suitable stimulating conditions, of an expression whose interpretation is that 'Fido barks' is true if and only if (or means that) Fido barks. And by extension nothing prohibits that the whole of our semantic capacity should depend upon having tacit beliefs of this sort.

Fodor's suggestion that knowing a language is a matter of translation of linguistic expressions into the language of thought, or as he puts it: "Learning English isn't learning a theory about what its sentences mean, it's learning how to associate sentences with the corresponding thoughts" (Fodor 1998: 9) is therefore additional both to the thesis that expressions are associated with thoughts, and that mind is representational. The key to Fodor's view is not the representational theory of mind, or the idea that words go with thoughts, but rather that learning English is a matter of "learning how".

As Fodor was among the first to argue at length, the notions of "learning how" and "knowing how" cannot be blandly accepted in psychology as though they had no intellectual component; and I find it ironical that Fodor, who is among those who pointed the way to a more suitable view of the relation between intellectual and practical abilities, should think of learning a language, but not, say, learning how to tie one's shoes, as a merely practical ability. But the question is whether learning and knowing a language are merely practical abilities or not, and the way is open to inquire further into the question whether, to paraphrase Fodor, learning English, or any other human language, is or is not learning a theory about what its sentences mean.

A point that is often not remarked in discussions of tacit belief, knowledge of a domain as theory construction, and so forth, is that there are many points where tacit and explicit knowledge interact. These invite interpretation in terms of theory construction. For example:

I first met Luigi Rizzi in Pisa, in 1979. Luigi spoke English pretty well, though with a marked Italian accent, and since my Italian was nonexistent we spoke in English. Luigi said, in a certain context, "Please remember me that", and of course I understood him, immediately and correctly, as intending to say would I please remind him of that. Here is a case where, detecting as I did an incorrect pattern of use, I traced it to a word incorrectly assumed by Luigi to be a causative in English (in fact, it is a causative in Italian). In Davidson's (1986) terminology, my "passing theory" of Luigi-interpretation so adjusted itself that I added 'remember' as a zero-affix causative to my assumptions about his lexicon. No doubt part of the explanation of my capacity to move to the right passing theory is that 'remember' does have a causative, namely 'remind', and there are plenty of zero-affix causatives in English.

The above illustration shows interpretive conjecture at work. The case is typical, in that conceptions drawn from linguistic theory (the zero-affix causative) interact with explicit beliefs to produce an interpretive outcome. As the literature on language acquisition attests, there are many cases where explanations are far more theoretical. In much of the literature there has been a great emphasis on the automatic character of first-language acquisition, and the speed and accuracy of perception. This emphasis is all very well in its way; but it discourages the thought that I am concerned to advance, that the ascription of meaning to linguistic objects is profitably seen as a matter of hypothesis formation, and that their production exemplifies a decision problem that it is part of our rational nature to engage in.

I have urged through simple example that tacit (or even unconscious) knowledge of the theoretical substance of linguistic structures is involved in the explanation of our interpretive capacities. My concern has been to outline the role that I see referential semantics, conceived as tacit knowledge of conditions on reference, should play in cognitive science: if I am right, referential semantics can aspire to deliver up the objects, apprehension of which through mental representations and their computational interaction is responsible for our semantic capacities. Such traffic as there may be between the theory of what is represented and the theory of its representations is a matter for another discussion.

Philosophical, rather than cognitive-scientific, issues about tacit knowledge have not formed part of my brief, but I will say a few words about them in closing. The realm of tacit knowledge, however scientifically cogent, seems to be nothing to us, because it cannot, by definition, participate in our conscious lives, and cannot be involved in justification and criticism of the use of language. Linguistic meaning, whether described in terms of reference or not, involves norms of speech, which Luigi Rizzi in my example inventively failed to observe. The reflection that tacit knowledge is beyond the realm of norms of interpretation is correct in a way: we do not hold people to norms that we do not apprehend. But to explain how we are able to observe the norms that we do apprehend we must, I believe, allow tacit correction of beliefs of which we are unaware. If so, then far from it being the case that the normativity of language puts semantic competence out of court, our capacity to have linguistic norms in the first place actually depends upon it.

11 Theories of Concepts: A Wider Task

Christopher Peacocke

There are many tasks for a theory of concepts which are suggested by the very nature of a concept. Those tasks include:

elucidating a particular concept's role in fixing reasons for making judgements;

characterizing its normative role more generally;

describing its relation to the level of reference and truth;

explaining the relation of all of the preceding to understanding and knowledge;

and of course much else. I would be the first to endorse the view that there is great illumination to be gained by carrying out such tasks. Here I am going to argue, as a substantive and no doubt controversial thesis, that there also exists a wider task, one whose execution is made possible by the construction of substantive theories of particular concepts. This wider task is one with epistemological, metaphysical, and psychological aspects. The task is that of addressing what I call the Integration Challenge.

In the first three sections of this chapter, I outline some metaphysical and epistemological aspects of the Integration Challenge. The views outlined here are expressed less dogmatically—or at least, at greater length—in my book *Being Known* (Peacocke 1999). These sections can serve as an introduction to the issues discussed there for those unfamiliar with that material. In the fourth section of this chapter I consider some ways in which the Integration Challenge bears upon psychology. Readers acquainted with *Being Known* should move straight to that section.

1. The Integration Challenge and a Form of Strategy for Meeting It

The Integration Challenge is the challenge of providing, for any given domain of thought, a simultaneously acceptable metaphysics and epistemology, and of showing them to be simultaneously acceptable. In almost all areas of

I thank the Leverhulme Trust for continuing support; the Philosophy Program of the Research School of Social Sciences of the Australian National University for the use of their excellent research facilities; Teresa McCormack for valuable comments on an earlier draft; and the participants at the Lisbon meeting for illuminating discussion, in particular Ned Block, Donald Davidson, Daniel Dennett, and John Searle. The text printed here follows closely that of the presentation to the Lisbon conference in 1998. Further elaboration of the position outlined in Sections 1–3 is found in my book *Being Known* (1999). In particular, a fuller exposition of the ideas in Section 1 would make use of the idea of taking the output of an informational faculty at face value.

thought, the problem arises of reconciling a plausible account of what is involved in the truth of statements in that area with a credible account of how we come to know those statements. The problem can take several different forms. Sometimes we may have a clear conception of the ways in which we ordinarily come to know statements in a given area, but be unable to provide any plausible account of any appropriate truth-conditions whose fulfilment could come to be known by these means. Alternatively, we may have a clear conception of what is involved in the truth of statements in a given area, but be unclear on how our actual methods for coming to know the statements could possibly get us as far as truth so conceived. In other cases, we may be unclear on both counts. In yet others, we may be clear about the nature of the truth-conditions, clear about our actual ways of coming to know, but unclear on how reconciliation is to be achieved.

This description is rather abstract, so let me give some examples of the Integration Challenge.

(*a*) *Necessity.*   We naturally think of statements about what is necessary, and what contingent, as objective. The most developed metaphysical account of modal truth, however, has been Lewis's modal realism, which treats other possible worlds as things of the same kind as the actual world (Lewis 1986). This arguably makes modal truth radically inaccessible. It must be tempting to give less exotic theories of the nature of modal truth. Some previous attempts have relied on notions of provability, and these have not even been extensionally correct. Others have embraced equally implausible expressivist, conventionalist, or other mind-dependent treatments of modality. Yet others have said that we must take modality as primitive, but have given no account of what it is to understand this primitive notion. We need a metaphysical treatment of modal truth which retains the objectivity of modal thought, but not at the price of inaccessibility. A good treatment of modal truth must ratify our basic methods of establishing modal truths as sound methods. Extant accounts of the metaphysical modalities have, for the most part, not met the Integration Challenge.

(*b*) *Knowledge of the Intentional Contents of Our Own Mental States.*   Thinkers ordinarily know the intentional contents of their beliefs and other propositional attitudes, without inference or reasoning, and without checking on their environmental relations. It would also be widely (though not universally) agreed that the conceptual components of these intentional contents are individuated in part by the complex environmental relations in which a thinker must stand if he is to possess those concepts. So we have the phenomenon of a thinker's having non-inferential, non-observational knowledge of a property he possesses, a property which nevertheless requires that he stand in certain external environmental relations. How is this possible? In this domain, the metaphysics of what is involved in the enjoyment of states with intentional content is relatively clear. We have a relatively good preliminary understanding of the explanatory role of these states, an understanding which makes it entirely natural to expect external individuation of intentional content. What we do not have, as yet, is a developed epistemology which meshes satisfactorily with this metaphysics. There have been important attempts to appeal to the self-verifying nature of some psychological self-ascriptions to resolve this problem, but they treat only a special case.[1] A general solution is needed, one which squares with what we already know about externalism and about self-knowledge. The example of knowledge of the contents of one's own intentional states also shows that one should not think that every case in which the Integration Challenge arises is one in which there is an apparent problem of inaccessibility. On the contrary, in this area it is the very accessibility of many of our propositional attitudes and their contents which is contributing to the existence of the Challenge.

(*c*) *The Past.*   The past is perhaps the most intuitively compelling domain in which we seem to have a conception of a kind of statement which can be true though quite undiscoverably so. The intuitive metaphysics for such statements is realistic. The question of how our past-tense thoughts can have a content which is capable of undiscoverable truth has still not, in my judgement, been fully answered. We need to develop a theory which answers this question, and does so in a way which still ratifies our ordinary methods of finding out about the past as capable of delivering knowledge of states of affairs of a kind which can, on occasion, obtain unverifiably.

It is certainly true that, in this particular area, versions of the Integration Challenge have often been driven by verificationist or constructivist theories of meaning. The Challenge has often been pressed by those tempted to some underlying idea of meaning as somehow fixed by rational responses to certain kinds of presented state of affairs. One can, though, ask what is involved in having our realistic conception, and require an answer which integrates with epistemology, without having verificationist tendencies. A satisfying response to the Integration Challenge for this area should also say what is right and what is wrong with the idea that meaning is fixed fundamentally by responses to presented state of affairs.

What then do theories of particular concepts have to do with the Integration Challenge? A starting point for elaborating the bearing of theories of concepts upon the Integration Challenge is a connection between one fundamental family of concepts and epistemology. This connection is made explicit in what I call the Linking Thesis.

The Linking Thesis. There is a class of concepts each member of which can be individuated in terms of the conditions for a thinker's *knowing*

---

[1] For pioneering discussion, see Burge (1988).

certain contents containing those concepts; and every concept is either such a concept, or is individuated ultimately in part by its relations to such concepts.

Several approaches to the theory of intentional content attempt to individuate concepts in terms of certain conditions under which contents containing a target concept are accepted or judged. What is distinctive of the Linking Thesis is the claim that there is a class of concepts for which the theory of individuation can proceed in terms not just of acceptance or judgement, but in terms of knowledge. According to the Linking Thesis, a concept in this class can be individuated as the one for which satisfaction of such-and-such conditions by the thinker suffices for knowledge of given contents containing the concept. If the Linking Thesis is right, such a concept is individuated not merely by some canonical role in the formation of judgements, but by its role in the acquisition of knowledge. I call such concepts *epistemically individuated.*

Not every concept is epistemically individuated. The theoretical concepts of the sciences are not. Nor, if this is a different case, is the concept of probability, for instance. Though there are certainly ways of coming to know that a theoretical concept is instantiated, knowledge of such ways cannot in general be written into the possession conditions for theoretical concepts. It takes hard thought, and sometimes extraordinary insight, to work out how, via experimentation and observation, one might come to know a certain theoretical hypothesis. The hypothesis can be understood long before the thinker works out how it might be known.

The central method of coming to know truths involving concepts which are not epistemically individuated is by abduction, inference to the best explanation. But knowledge of truths involving concepts for which abduction is central to their epistemology is possible only if there are concepts for which abduction is not thus central. Abduction can yield knowledge of its conclusions only if the propositions explained in the abduction are themselves knowledge. If knowledge can be reached by abduction, there must be some things not known by abduction, on pain of regress. It is not possible for everything to be known by abduction, any more than it is possible for everything to be known by deductive inference. The impossibility has the same explanation in both cases. If a method is a method of coming to know only if certain propositions are already taken as known—a condition which is met for both abduction and deductive inference—then not all knowledge can be reached by that method. On the position for which I am arguing, knowledge involving epistemically individuated concepts is not so reached.

Let us return to considering that class of epistemically individuated concepts further. One way to make it plausible that the class is non-empty is to offer examples. A possession condition for an observational concept *F* plausibly entails that a thinker who possesses that concept will be prepared to judge that something falls under it when he has an experience of a certain kind, and is taking experience at face value. Let us call the conditions under which a given subject's perceptual faculty delivers experiences which are genuine perceptions its "informational conditions". (This could be refined, but it is enough to make the relevant points.) We can then formulate an epistemically oriented possession condition for the observational concept *F* in question. If the thinker has an experience of the appropriate kind; if he takes this experience at face value; if the informational conditions for his perceptual systems are fulfilled; if any presuppositions he is making are fulfilled; and if there are no other reasons for doubt: then our thinker will not only be prepared to judge, but will be in a position to know, a content of the form 'That is *F*'. Similarly, consider the logical concept of conjunction. If a thinker has the concept of conjunction, he must be willing to make the transition from the contents *p* and the content *q* to the conjunction *p&q*, and conversely. This seems to be required for possession of the concept of conjunction. If in addition our thinker knows the premisses of such a transition, the judgement of the resulting conjunction amounts to knowledge. The concept of conjunction could be individuated as one such that knowledge of *p&q* is attained when that conclusion is inferred from known premiss *p* and known premiss *q*; and such that knowledge of each of *p* and *q* separately is attained when each is inferred from the known premiss *p&q*. And so on: one could continue to look at credible philosophical accounts of the possession of various concepts, and case-by-case one could endeavour to confirm that these accounts could be transformed into accounts which speak of knowledge of certain contents containing the target concept. One could, case-by-case, build up a specification of a class of concepts of which it is true that its members are epistemically individuated.

Pursuit of this approach will, however, yield rather limited philosophical illumination. If there is a class of concepts which conforms to the first part of the Linking Thesis, we will want not merely an enumerative characterization of its members. It would be good to have some more fundamental characterization which explains why any given concept which is epistemically individuated is so.

I suggest that any concept which can be individuated in terms of the conditions for accepting or judging certain contents containing it will also be an epistemically individuated concept. This generalization is confirmed in the examples already given. We can also offer a four-step argument for the generalization. The argument merits a lot of further elaboration, but here is the essence of the case, with an outline of supporting reasons.

(1) The judgements mentioned in the possession condition for a given, target concept which is individuated in terms of its role in judgement cannot be merely rationally discretionary judgements, rationally optional for the thinker. They must be rationally non-discretionary judgements.

This first step is no more than a spelling-out of what is involved in something's being a possession-condition for a concept. If judgement of a content can be rationally withheld, in given circumstances, then it cannot be part of the possession-condition for any of the concepts in the content that the judgement be made in those circumstances, cannot be something the possessor of the concept is required to judge.

(2) Rationally non-discretionary judgements aim at knowledge.

This second step is supported by the consideration that the thinker who is making what he takes to be a rationally non-discretionary judgement must, other things equal, withdraw it if he comes to accept that a judgement made in his circumstances would not be knowledge. If you come to accept that, had you been looking in a different direction, you would be seeing a hologram of a coin, your actual judgement 'That's a coin', based on your perceptual experience, must either be withdrawn, or given some new grounds. The presuppositions which made it rationally non-discretionary have been undermined.

(3) So if the suitably attained presuppositions of the thinker when making a rationally non-discretionary judgement are fulfilled, and any beliefs on which he is relying in making it are knowledge, the rationally non-discretionary judgement will be knowledge.

If the rationally non-discretionary judgement were not knowledge when all the thinker's presuppositions are fulfilled, and any beliefs on which he is relying are also knowledge, then appreciation of this insufficiency could rationally lead a thinker rationally to withhold the judgement. So the judgement would not after all be rationally non-discretionary in these circumstances.

(4) Hence the target concept could be individuated in the following way. It is that concept which, when certain judgements involving it are made by specified methods; and when any properly made presuppositions are fulfilled; and when the informational conditions for any faculties involved are also met; and when any beliefs on which the thinker is relying are knowledge, then: the judgements so reached involving the concept are knowledge. Hence, two different epistemically individuated concepts will involve at some point or other either different presuppositions, or different beliefs relied upon, or different methods in the conditions for corresponding judgements involving the two concepts to be knowledge.

This general approach may make it seem as if rationally non-discretionary judgement is something very strong, and in the nature of the case a rare occurrence. On the contrary, I think it is commonplace. In a wide range of cases, judgements made on the basis of evidence or reasons which may seem inconclusive, and so may seem rationally discretionary, are in fact cases in which, when one takes into account the thinker's presuppositions and background beliefs, judgement is rationally non-discretionary. Visual experience as of a phone in front of oneself does not make the judgement "That's a phone" rationally non-discretionary. But in ordinary circumstances the thinker presupposes all of the following: that he is perceiving properly; that objects which have one side like that of a phone have the remainder of the shape and material properties of a phone; and that objects of a certain shape and set of material properties have the function of permitting long-distance conversation by a certain means. In the presence of these presuppositions on the part of the thinker, the judgement "That's a phone" becomes rationally non-discretionary. That each of these is indeed a presupposition is evidenced in part by the fact that the thinker is rationally obliged to reassess his judgement "That's a phone" were he to come to believe that any one of them failed to hold. Similar points apply to judgements based on the deliverances of experiential memory, propositional memory, and testimony.

The conclusion of the argument (1)–(4) is not that there is, for concepts individuated in terms of outright judgements involving them, an additional layer of epistemic requirements in addition to those formulated in terms of judgements. The argument is rather that when we appreciate the rationally non-discretionary nature of the judgements mentioned in such possession conditions, we are in a position to develop an argument that the possession conditions for such concepts can equally be formulated in terms of knowledge. The point is not that the formulation of the possession condition in terms of judgement is in some way incorrect: it is not. The point is that the formulation in terms of knowledge is equally available.

This argument has various consequences for the way in which we should conceive of the relation between concepts and knowledge. Our present task is to consider one of them: the way in which the Linking Thesis leads to a strategy for addressing the Integration Challenge in one range of cases.

I will be working within the scope of the background supposition that a substantive theory of intentional content must determine an attribution of truth conditions to intentional contents (at least in fundamental cases). Under this supposition, a theory of a specific concept will determine a contribution to the truth conditions of intentional contents in which it features. Intentional contents are here conceived as lying at the level of sense. Under this background supposition, the level of sense is inextricably and fundamentally involved with the level of reference. Senses fix truth-conditions, which must be characterized via the level of reference. If senses fix truth-conditions, it must be possible to individuate a sense by giving the condition for something to be its reference.[2] Several different styles of

---

[2] This principle differs by only a hair's breadth from one Dummett has long emphasized: see e.g. Dummett (1993). The hair's breadth difference is the difference, not relevant here, between formulations which do, and formulations which do not, give philosophical priority to language.

substantive theory can succeed in determining an assignment of truth-conditions, if they are related in the right way to the level of reference. Various species of conceptual-role theories, theories of implicit conceptions and information-based semantics can each, in their own way, do so.

The significance of the Linking Thesis for the integration of metaphysics and epistemology can then be formulated as follows. One way of meeting the Integration Challenge for a given domain is to supply a substantive theory of content for the concepts of that domain. By the Linking Thesis, the theory of intentional content can in those cases be framed in terms of knowledge, in the way we described. The Integration Challenge is that of showing how the methods by which we normally think we come to know contents of a given kind really do ensure the holding of the truth conditions of those contents. This challenge is answered head-on if we can develop a theory of the concepts of the domain under which a judgement which is rationally non-discretionarily required for possession of a concept of the domain must be both true and knowledgeable when its presuppositions and informational conditions are fulfilled, and its premises known. A good theory of content will, if the Linking Thesis is correct, close the apparent gap that led to the Integration Challenge in cases in which a solution relies fundamentally on truth-conditions, and in which the relevant concepts of the domain are epistemically individuated.

This claim does not involve some horrible, illegitimate slide from the level of sense to the level of reference and metaphysics. A substantive theory of a concept is indeed something concerned with the level of sense. But if a theory of concepts or sense must determine an assignment of truth-conditions, we already have a connection with the level of reference. On the position for which I have argued, the theory of grasp of the concept can be formulated in terms of knowledge of certain contents; and knowledge of the correctness of a content requires the world to be a certain way. So there is no illegitimate slide. Rather, there is in these cases a general connection between the conceptual, the epistemic, and the metaphysical.

On this strategy for addressing the Integration Challenge in the cases meeting the two conditions, the Janus-faced character of the theory of under-standing is pivotal. The theory of understanding has both a metaphysical aspect, in being a theory of grasp of truth-conditions, and an epistemic aspect, having to do with rationality of judgement. This epistemic aspect, when we are concerned with epistemically individuated concepts, has to do not only with justification, but with knowledge. To have a rift between our epistemol-ogy and our metaphysics is, in these cases, necessarily to have some fragmen-tation in our theory of understanding and concept-possession.

This formulation of the significance of the Linking Thesis should make it very clear that the importance of the Thesis lies in determining *what* has to be done to meet the integration challenge in any given area. What we have to do is to provide a substantive theory of concepts in that area, a theory which can be cast in terms of conditions for knowing certain contents involving those concepts. This formulation of the significance of the Linking Thesis does not specify *how* to write out such theories in the philosophically interesting cases. All the hard work of meeting, for any given philosophically interesting area of thought, the goal fixed by the Linking Thesis has still to be done. We have specified the form of a solution, but not yet its content.

2. Three Indicators for Solutions

How then are we to make progress in identifying what kind of substantive solution to the integration problem might be appropriate for a given area of discourse meeting our two conditions? We can start by asking: are there any features of truth in a given area which can provide guidance about the kind of theory of content for that area which would, via the Linking Thesis, help us to meet the integration challenge?

We can proceed by developing a set of indicators. An indicator is a property whose instantiation, or non-instantiation, by statements of the area in question ought prima facie to be explained by a good integrationist solution. The presence or absence of an indicator in a given area thus constrains philosophical treatments of that area, and thereby offers some guidance in constructing integrationist solutions. An indicator might be thought of as a light, which is either on or off for a given area. To fix ideas, I will consider indicators which are particularly pertinent to the modal and temporal cases, and discuss their application in those areas. The indicators do, though, also apply to other areas. The first indicator is given in the question:

(1) Do true statements in the area have an a priori source?

By 'having an a priori source', I mean the following. In the modal case, we know from the work of Kripke that not all true statements of metaphysical necessity are a priori (Kripke 1980). 'Necessarily, if Hesperus exists, Hesperus is Phosphorus', and 'Necessarily, water is constituted by H_2O' are familiar examples. But in both of these cases, the true a posteriori necessities are consequences of two premisses, neither of which is both modal and a posteriori. One of the two premisses is modal but also a priori; while the other premiss is, though a posteriori, non-modal. In the case of 'Necessarily, if Hesperus exists, Hesperus is Phosphorus', the a priori premiss is the necessity of identity. The a posteriori, non-modal premiss is that Hesperus is Phosphorus. In the case of the statement about water, we have a similar division into the a priori premiss that necessarily, water has its actual constitution, and the a posteriori non-modal premiss that water is constituted by two parts of hydrogen and one of oxygen. I suggest that such a tracing back to a priori sources is always possible. Any a posteriori necessity rests, it seems, ultimately on principles each of which is either modal and a priori, or a posteriori and also non-modal. If correct, this is something which should be explained by any integrationist solution.

This is in sharp contrast with the case of the past. Consider the statement 'It rained in Los Angeles yesterday'. It is quite implausible to suggest that this statement involving the past is a consequence of two other statements, one involving the past but a priori true, and the other not involving the past at all. True statements about the past are not somehow determined as true on the basis of some a priori principle or principles which, when taken in conjunction with other past-free a posteriori statements, fix the truth-values of statements about the past. Statements about what was the case at particular past times are just brute truths or brute falsehoods. There is no a priori determination of past-tense truths by other truths not about the past. A solution to the integration problem for the past must explain how we have such an understanding of the past, and why the case is so different from that of modality.

The second indicator involves causation, and is given in this question:

(2) Is some role in causal explanation essential either to the truth of statements in the area in question, or to our having our concepts of that area?

The answer to both parts of the question is negative in the modal case. Only what is actually so can enter causal explanations. The fact that something is necessarily so, or possibly so, never causally explains anything. Apparent counterexamples to this principle are cases in which it is someone's propositional attitudes to modality, or proofs about modality, or more generally some operator concerned only with what is the case in the actual world, which is involved in the causal explanation. If indeed the modal does not enter into causal explanation, we may correspondingly say in Humean terms that there is no impression from which the idea of metaphysical necessity is derived. This is partly what has made the epistemology of modality so problematic.

This second indicator in the modal case can also serve to illustrate the point made earlier, that the Integration Challenge can also be formulated as a problem about the nature of understanding. It is very tempting to say that operators like 'necessarily' and 'possibly' should be treated as primitive. This is certainly attractive when we compare the position with that of the modal realist who takes these operators as quantifiers over worlds of the same kind as the actual world. Taking the operators as primitive, however, should be accompanied by some account of what it is to understand these primitive operators. In other cases where we have primitive predicates which are understood, such as observational predicates, the account of understanding involves some causal interaction between instances of the property picked out by the predicate and the understander's use of the predicate. The second indicator in the modal case implies, correctly, that this account of understanding a primitive predicate or operator is not available in the modal case.

By contrast, I would argue that the capacity of temporal relations to enter causal explanations is an essential component of an account of our mastery of temporal relations, including our capacity to think about the past. I will return to this when I sketch a response to the Integration Challenge for discourse about the past. For the moment, I note that we would expect certain answers to the question defining the second indicator to go together with certain answers to the question defining the first indicator. When an area is involved in causal explanation, faculties or devices which are causally sensitive to states of affairs in that area will provide a means of obtaining knowledge of some of the truths about that area. Knowledge obtained in that way will not be a priori. So, for a given area, a positive answer to our second question about a constitutive role for causality will naturally be accompanied by a negative answer to our first question about the a priori. Conversely, a positive answer to the first question for a given area should be expected to imply a negative answer to the second question. If all knowledge in the domain in question can be obtained by a priori methods, causal interaction with the domain cannot be essential to thought or knowledge about it. So, the two indicator lights fixed by the first questions are, for a given area, both on together, or both off together.

The third indicator I want to mention involves an identity of property, and is given in this question:

(3) Are statements in the problematic area predications of a property which also features in predications outside the problematic area?

If the answer to this question is affirmative, and we have some understanding of mention of the property outside the problematic area, this understanding can provide some constraints on both the metaphysics and the epistemology of statements in the problematic area.

In the case of the past, there are such identities of properties. For instance, we can truly say this:

The thought (or utterance) 'Yesterday, it rained' is true if and only if yesterday had the same property as today is required to have for a present-tense thought (or utterance) 'It is now raining' to be true.

We could equally have written "if and only if yesterday had the same property as any arbitrary day is required to have for a present-tense thought (or sentence) 'It is now raining' to be true with respect to that day".[3]

---

[3] If the language uses temporal operators, rather than any ontology of times, the issue of whether there is a property-identity must be addressed at the level of the corresponding statement involving an ontology of times. The question raised by the third indicator cannot be answered in the negative simply by using a system of operators rather than an ontology of times. An alternative way of meeting the need would be to expand the identities in question to include not only properties, but also the identity of the way it has to be today for 'Today——' to be true with the way it had to be yesterday for 'Yesterday——' to be true.

Appreciation of this property-identity clearly cannot amount by itself to understanding 'yesterday'. The right-hand side of the biconditional simply uses the past tense, so any kind of appreciation of the biconditional must presuppose some possession of the concept *yesterday*. Nonetheless, this property-identity is a very substantial constraint upon the metaphysics and epistemology of the past. No account which is inconsistent with it can be acceptable; and we ought to aim for some account which explains why it is true. The property-identity also has some explanatory power. Suppose yesterday you learned something which you would then have expressed by saying 'It is raining now', and that you store what you have learned in memory. If, today, you want to express what you learned yesterday, saying 'Yesterday it was raining' is quite sufficient: you do *not* have to realign or re-adjust the predicate which was yesterday combined with the word 'Now'. The property-identity explains why you do not have to. If yesterday did indeed have the property required for 'Now it's raining' to be true with respect to it, which is what you learned, then by the property-identity that is sufficient for 'Yesterday it rained' to be true.

Property-identity principles are far from being trivialities. When we consider this third indicator in relation to certain treatments of the modal case, we have an example in which there is a failure of certain property-identities. Suppose the actual world is conceived of in David Lewis's way, as you and all your surroundings (Lewis 1986: 1). Then consider this biconditional:

> The thought 'It is raining in Oxford at the start of the twentieth century' is true with respect to a given possible world if and only if that world has the same property the actual world is required to have for the thought 'It is raining in Oxford at the start of the twentieth century' to be true.

This biconditional seems to me to be one we must reject, unless we are Lewisian modal realists. If a possible world is conceived of as a set of propositions, or sentences, or suppositions, *and* if the actual world is not so conceived, then the property the actual world has when it is raining in Oxford at the start of the twentieth century is one which it is impossible for a possible world, so conceived, to have at all. The property of having rain falling in one of its spatio-temporal regions is not a property a set of propositions, or sentences, or suppositions, could ever have. What such a set can have is the property of being such that *according to it*, it is raining in Oxford then. But the property

x represents rain as occurring in a certain spatio-temporal region

is sharply to be distinguished from the property

x has rain occurring in a certain one of *x*'s spatio-temporal regions.[4]

This third indicator is also closely linked to the second. If we have a causal epistemology for a certain area, and properties in that area are identical with those predicated in the problematic area, it would be quite puzzling if there were not a causal epistemology in the problematic area. Similarly, contraposing, we would not expect there to be property-identities between a problematic area whose epistemology is non-causal and one whose epistemology is causal. Since I also noted that the second and first indicators can be expected to be both on or both off for a given area, we now have reason to expect that, for any given area, these three indicator lights are all on together for that area, or all off together. If what I have been saying so far about the modal and temporal cases is right, they both conform to this expectation of correlation between the indicators.

3. Two Styles of Solution

I now turn to outline two radically different models of ways in which integration may be achieved. Each model is capable of explaining a pattern of indicator lights. I will outline one model as it can be developed for the case of the past tense, and the other as it can be developed for metaphysical necessity. The models are just two of several integrationist models which must exist. Certainly some other subject-matters must be treated rather differently from those which do fit the two models I will be discussing. The two models do, though, serve to illustrate two paradigmatically different ways in which the Integration Challenge can be met.

The first model can be called *the model of constitutively causally sensitive conceptions*. The outlines of this model are already strongly suggested by cases in which the indicator of constitutive causal connections is lit up. On this model, one part of an account of grasp of temporal concepts such as *earlier than* is a causal sensitivity of certain judgements involving it to instances of that very relational concept. Temporal relations themselves, and not just evidence for temporal relations, are capable of entering into causal explanations. They do so when you knowledgeably remember one event as occurring before another. They do so even when you remember participating in a certain kind of event. The fact that such an event occurred and was prior to the present time causally explains your impression that it occurred before

---

[4] Parallelism can be restored if the actual world is conceived of as the set of true propositions, sentences, or Thoughts, and non-actual worlds are conceived of in the way in which Lewis describes as "ersatz". The property–identity link this generates, however, cannot be used in a theory of modal understanding, since it already presupposes a division between those sets which represent genuine possibilities, and those which do not. Grasp of this distinction is what has to be explained by a theory of understanding.

now. This model embraces a form of externalism about certain temporal concepts. What it is for an impression to have a past-tense temporal content is in part a matter of its being explained, in central cases when all is working properly, by certain temporal relations themselves. Such an externalism also explains an ordinary thinker's entitlement to take temporal impressions at face value, in the absence of positive reasons for doubt.

Such a constitutive causal sensitivity to temporal relations themselves must be embedded in a conception of the past as the past. The model can also be developed in a form which embraces the view that one who understands the past tense has tacit knowledge of the temporal property-identities we noted earlier. The further development the model needs has to include elaboration of how this aspect of understanding is intertwined with the externalism.

Even from this simple sketch, though, it appears that the model of constitutively causally sensitive conceptions promises an explanation of the pattern of indicator lights displayed by the case of thought about the past. Truths about the past can hardly be fundamentally a priori if the fundamental way of coming to know them involves the exercise of faculties causally sensitive to temporal relations and properties. Grasp of the property-identities is also part of the model, as I just described it. That is, the same identities which are involved in the truth of a past-tense thought are also involved in grasping it. This is a feature we will find in other models too.[5]

The other model is the model of *implicitly known principles*, and it seems to me the appropriate model for meeting the Integration Challenge in the case of metaphysical necessity. What makes a set of sentences, or propositions, or thoughts into a genuine possibility? The model of implicitly known principles, as I would develop it for the modal case, claims that there is a set of principles, "principles of possibility", which any set of sentences, thoughts, or propositions must satisfy if it is to represent a genuine possibility. To understand modal discourse is to have implicit knowledge of these principles of possibility, and to deploy them in evaluating modal sentences.

The most fundamental of these principles of possibility states that some putatively possible world-description w is a genuine possibility only if the world-description involves the application of the same rule in w to determine the extension of the concept in w as is applied in determining its extension in the actual world. Any world-description which describes the conjunction A&B as true, but A as false, or a person as a bachelor but not a man, will not be applying the same rules for determining the semantic values of conjunction, or the predicate 'is a bachelor', as are applied in determining their extensions in the actual world. Other principles of possibility will deal with the preservation in genuinely possible worlds of what is constitutive of objects, properties, and relations. There will also be a second-order principle

---

[5] My own view is that appreciation of the import of these identities rules out any constructivist or evidential account of past-tense truth. The theme is further developed in ch. 2 of *Being Known*.

of possibility, to the effect that something is possible if it is not ruled out by the other principles of possibility. According to this approach, 'It is metaphysically necessary that p' is true if and only if p is true in every world-description which respects the stated principles of possibility.

If these principles of possibility are a priori, and they succeed, in combination with non-modal truths, in fixing modal truth, then we would have an explanation of why modal truth has, in the sense identified earlier, a fundamentally a priori character. On the model of implicitly known principles, no causal contact with some modal realm is required for understanding. Nor, on this model, does modal truth involve any realm of worlds of the same kind as the actual universe. The failure of the property-identity principle for the case of metaphysical necessity is implicit in the background framework, should it prove adequate to modal truth and understanding, since that framework treats non-actual possible worlds as sets of sentences, propositions, or thoughts—that is, not as things of the same kind as the actual world, as the property-identity principle requires. In these ways, the model of implicitly known principles is capable of explaining the pattern of indicator lights for the case of metaphysical necessity.

4. Psychological Aspects of the Integration Challenge

Meeting the Integration Challenge is not only a task for philosophy. It is also a task for psychology. It would be widely agreed that the question 'How are humans able to think such-and-such contents?' has both philosophical and psychological aspects. Exactly the same holds of the question 'How are humans able to *know* such-and-such contents?' A philosophical answer to this question about knowledge generates a distinctive set of psychological questions.

A philosophical answer to the Integration Challenge in a given domain will characteristically identify certain properties of understanding which are involved in the capacity to think about that domain. The philosophical answer will elaborate some of the relations of that understanding to both the metaphysics and the epistemology of that domain. Thus, in the modal case, we identified tacit knowledge of certain principles of possibility as constitutive of modal understanding. These principles of possibility are linked both to truth, in fixing the truth-conditions of modal contents, and to knowledge, in that their correct use can lead to modal knowledge. In the temporal case, we identified certain abilities to be responsive to temporal relations themselves as an essential component of temporal thought and knowledge. The mental states and capacities identified in philosophical answers to the Integration Challenge immediately raise a set of questions for psychology. Perhaps the most salient of these is a "how"-question: How are humans capable of being in mental states which have the properties and relations identified in these philosophical accounts?

The question I intend here is not a question about acquisition, nor about the evolution of mental states, important and interesting though those questions are. It is a question about the realization and present relations of the attained state, whatever its history of acquisition. Evolutionary psychologists and epistemologists sometimes argue that humans, or some other species, would not survive unless certain basic methods of belief-formation employed in that species were generally correct; or were knowledge-yielding; or unless certain kinds of concept-formation take place. Whether or not these claims are correct, their proponents take for granted that humans or members of whatever other species is in question are capable of being in certain attained states, with certain kinds of contents. They take for granted—with entitlement, for their purposes— that there is some answer to the question of how the subpersonal states of a thinker can be realized in such a way as to support the properties and relations required for a kind of understanding which can meet the Integration Challenge. But the "how"-question cannot be postponed indefinitely.

An illustration of a satisfying form of answer to this sort of how-question is provided by the case of perception. Any philosophical account of how humans have knowledge about their immediate environment will identify the existence of perceptual states which play a part in the generation of this knowledge. The philosophical account of such knowledge will require the perceptual states which lead to knowledge to have contents about the perceiver's surroundings which are correct, in respect of those contents which yield perceptual knowledge. A subpersonal psychological theory of how the content of the visual experience is computed from the proximal stimulus, by means which reliably generate correctness, will provide a satisfying answer to the question of how the perceptual state can stand in the relations required if it is to play its part in the generation of knowledge.

A computational theory of vision is a genuinely empirical account. It is also not a theory of how an organism acquires the capacity to have perceptual states. The same computational capacity might have been acquired or evolved in several different ways. The computational theory is rather a theory of the relations in which various subperson states stand to one another, to the environment, to the proximal stimulus, and to the visual experience, once such a system is acquired. For each of the mental states identified in a plausible response to the Integration Challenge in a given domain, we need an equally satisfying psychological explanation of how humans can enjoy them, and how they can stand in the required relations.

It is evidently not a purely philosophical task to develop such theories. There are, though, certain constraints of a relatively general, constitutive nature that any good psychological account must respect. To say that there are such constraints is not to engage in philosophical imperialism, is not to say that philosophical claims have some weight which empirical discoveries lack. On the contrary: it is evident that psychological results can show some constitutive proposals to be false. My point is just that constitutive constraints must be respected. Positive or negative evidence about what these constraints are may in principle come from many different sources. Saying that a theory is not respecting, or not explaining in accordance with, some constitutive constraint is no more philosophical imperialism than it is mathematical imperialism if a mathematician finds a mathematical error in a physical theory. Any theorist should, for instance, acknowledge that a psychological theory which does not explain why our perceptions are as reliable as they are cannot be an answer to the question of how humans are capable of knowing as much as they do. It is not rationally open for someone to defend such a psychological theory by saying, 'Well then, perceptual knowledge does not require reliability.'

In illustration of this point, we might also imagine a subpersonal theory which simply takes it for granted that certain states have a reliably correct informational content, and uses this assumption in trying to explain the reliability of perceptual states. Such a theory would be taking for granted part of what needs explanation. No such theory could do the full explanatory work required of a psychology of perception.

My aim in the remainder of this chapter is to identify two ways in which a psychological theory may not succeed in characterizing psychological states which have the properties and relations identified as required in a solution to the Integration Challenge. The ways can be generically labelled as those of *insufficiency* and *non-necessity*. These are not the only possible ways in which a theory may fall short in relation to the Integration Challenge. Nor is avoiding them the only possible source of strength. I have chosen them here partly because they are so clearly related to the preceding philosophical discussion, and partly because they can be illustrated by interesting and significant proposals actually adumbrated in the psychological literature. For these purposes, I will also select domains for which we have already considered the shape which ought to be taken by a philosophical answer to the Integration Challenge.

Insufficiency

Philip Johnson-Laird's paper "The Meaning of Modality" (Johnson-Laird 1978) contains what is, to the best of my knowledge, the only attempt of its time to develop a psychological theory of the nature of modal understanding. His aim is to "elucidate the mental representations of the meaning of modals" (p.18). He describes a procedure for evaluating modal statements that involves "mentally constructing a sequel" to some "reference situation". His view is that modality has a uniform meaning, but that in attempting to construct these mental models, we sometimes use 'deontic knowledge' and sometimes other bodies of information. The statement 'It is possible for John to leave' is one which may be understood deontically. Of the deontic knowledge on which we may draw in evaluating it, Johnson-Laird writes that "the rules of games, the conventions of society, and the laws of the land are typical examples of bodies of deontic knowledge" (p.19). Johnson-Laird contrasts this interpretation with what he calls the epistemic interpretation, for the evaluation of which we draw on what he calls epistemic knowledge. He writes: "The naïve physics and psychology of everyday life, as well as their sophisticated scientific counterparts, are typical examples of epistemic knowledge" (p.19).

If we try to apply this approach to understanding of metaphysical possibility and necessity, presumably we would correspondingly say that, in evaluating modals so interpreted, we draw on our body of absolute modal knowledge. But so far from giving us an account of the mental representation of metaphysical modality, this approach seems to presuppose that problem is already solved. For a thinker's modal knowledge must either contain modal notions in its content; or else non-modal contents must be stored in some form in a functional location labelled with something which means "The Necessary". If there is no such modal element, either in the content or the labelling of the storage, the account will not work. Not just any information can be used in evaluating modals. So the approach is not by itself sufficient to solve the problem of how modality is mentally represented. Johnson-Laird's theory is a psychological analogue of a philosophical theory which says that a statement 'It is possible that *p*' is true iff *p* is consistent with the laws of modality. This biconditional is true enough; but without some account of what it is to appreciate something as being a law of modality—which looks like the same problem of understanding modality again—then it is not sufficient to meet the explanatory need.

This point is not restricted to metaphysical modality. The point is essentially of a structural nature, and it seems to me to apply to each of the interpretations of modality Johnson-Laird mentions. It may be fine to evaluate a statement "It is possible for John to leave", under an interpretation concerned with what is legally possible in a given country, by trying mentally to construct a sequel to some reference situation, where the sequel is constrained to conform to what is legally required by the laws of the country in question. This is just to push back the question from understanding of legal possibility to that of legal requirement. The understanding and mental representation of the notion of a legal requirement remains unelucidated on this treatment. The corresponding point could be made for any treatment of any interpretation of the modality cast in this mould.

The insufficiency would be avoided if we could give some further substantive elucidation of what it is to understand the necessity in question. In the case of metaphysical necessity, of course, I would say that such an account could be given by mentioning the thinker's tacit knowledge that this notion conforms to what I called the principles of possibility. Such a principle-based approach could be married to an approach to the mental representation of modality which involves the construction of mental models. So what I am saying here is not in itself an objection to the mental models approach. My point at the moment is just that some much more substantive theory is needed, beyond that in Johnson-Laird's 1978 paper, if we are to avoid the insufficiency problem. Indeed it is only with such a more substantive theory that one could start on the project of explaining how not merely modal belief, but modal knowledge, is psychologically possible.

Non-necessity or overascription

As an example of what I will argue is a kind of overascription, when considered in relation to the account of temporal understanding I outlined, I will consider the striking theory of Josef Perner and Ted Ruffman in their paper "Episodic Memory and Autonoetic Consciousness: Developmental Evidence and a Theory of Childhood Amnesia" (Perner and Ruffman 1995). Perner and Ruffman offer a theory of what is involved in having an episodic memory of an event, and aim to explain various empirical phenomena by means of this theory. To remind ourselves, episodic memories are characteristically described in such sentences as 'I remember watching the results of the 1997 General Election come through on television.' Episodic memory is a matter of remembering seeing, experiencing, thinking, doing...on some particular occasion in the past. Episodic memory is contrasted with what psychologists call 'semantic memory', the kind of memory you have that the Battle of Hastings was in 1066. I remember that I was born in Birmingham (semantic memory), though of course I do not remember being born (no episodic memory). It is sometimes regarded as a characteristic of episodic memory that the rememberer of the episode remembers or knows when and where the episode occurred. This is too strong for episodic memory as understood here. You can remember having a conversation with someone, while wondering whether it took place in Oxford or in London, and whether it took place last summer or the one before. Conversely, semantic memory is sometimes mischaracterized as knowledge of general truths. Though if general information is remembered, it must be a case of semantic memory, all the above examples of semantic memory are memories of particular truths. Equally, memory that there was once and only once an event of such-and-such kind in which one participated is not enough for remembering it, as the case of memory that one was born at a certain place and time shows.

Perner and Ruffman's hypothesis—henceforth, simply "the Hypothesis"— is that having an episodic memory involves the rememberer representing to himself (*a*) that he has experienced the event in question before, and (*b*) that he knows about the event now because he experienced it in the past. Thus they write:

For merely *knowing* that there was the word "butter" on the list it is sufficient to represent the proposition *there was the word "butter" on the list*. In contrast, for an act of proper *remembering*, in this specific narrow sense, it is also necessary to represent the additional fact that *one has seen the word on the list and that, therefore, one knows that the word has appeared on the list.* In other words, one has to represent one's own act of experiencing. (1995: 517)

There is relevant developmental work suggesting that children may lack the prerequisite representations for recollective experiences until they are 3 to 5 years old. This is not to say that children before that age may not, in Tulving's words, *know* something about the past as they certainly do. For instance, Fivush and Hamond (1990) have shown that even at 2 children can retrieve much detail about a trip to the zoo. Such demonstrations of early competence, however, do not show that children have recollective experiences of *remembering* these past events. Such recollective experiences may not be had before the age of 3 or 5 years, because children up to that age do not understand the connection between informational access (e.g., seeing) and knowledge (Wimmer, Hogrefe, and Perner 1988; Wimmer, Hogrefe, and Sodian, 1988). (1995: 517–18)

Amongst the empirical phenomena that Perner and Ruffman aim to explain by the Hypothesis is the phenomenon of childhood amnesia, the phenomenon that most people are unable to remember their early childhood.

So the main contribution of this research to explaining childhood amnesia is to explain adults' difficulty *remembering* (in the strict sense) their earlier life history. They cannot *remember* because without having encoded early events as experienced they cannot have the phenomenal recollective experience of having experienced these events, though, and this is very important to note, they may *know* something about these events. (1995: 542)

I raise two problems for the Hypothesis. One concerns overascription. The other is a dilemma which arises for the content of the representations Perner and Ruffman describe.

Problem One   It seems that someone can remember seeing something without having the concept of seeing or of experience. In describing the memory as one of seeing, we characterize the episodic memory as one of a certain subjective type, which is phenomenologically different from remembering hearing, remembering tasting, remembering feeling. The memory experience can be of one of these distinctive types without the rememberer having concepts of these types.

Equally, within the representational content of the episodic memory itself, there need not be any use of the concept of experience or perception. The memory may represent oneself as once travelling on a train through Scottish countryside, with such-and-such an array of objects around one and so-and-so movements and events occurring. This representational content no more has to involve the concept of experience, or experience of a certain kind, than does the original visual experience which (when all is working properly) produces it. This is only to be expected, since the content of the episodic memory, again when all is functioning properly, is contained in, or at least in context determined by, the content of the original experience.[6]

To enjoy an episodic memory of an event without the rememberer exercising any concept of experience is quite different from having merely semantic memory that one was involved in an event of a certain kind. Only the episodic memory can be described as a remembering seeing, or hearing, or feeling ... If children below the age of about three-and-a-half years do not have the concept of experience, that does not preclude them from having episodic memories, since having such memories does not require possession of the concept of experience. Perhaps the suggestion is not that young children do not have the concept of experience, but only that they fail to know that experiences are a source of knowledge. But it equally seems that a person can have an episodic memory of seeing something without having beliefs about the relations between experience and knowledge.

It is, of course, not sufficient for something to be an episodic memory that it is a conscious state with an imagistic content produced by the operation of memory. Imagistic memory representations could be generic, and even part of a script.[7] That is different from an episodic memory, which rather represents for the subject specific historical episodes in the subject's own past. Genuine episodic memory is possible only for a subject who has some conception of a spatio-temporal world, and for whom the subjective experience of remembering something represents some specific historical state of affairs or events in which he participated, and which occurred at a particular place and time (even if he does not know when and where it was). This spatio-temporal conception must be in place for genuine episodic memory to be possible. Possessing this conception involves integrating the contents of perceptual and memory states into a conception of the spatial layout and the temporal development of the world the subject encounters. It is one thing to be able to do that; it is another to reflect on it, to have the conceptual resources to describe what it is one is doing. Though the matter needs further argument and development, it seems to me that there is no sound argument from the agreed necessity of this background conception of an objective world to the conclusion that the subject must himself have the concept of experience, and must have and exercise correct thoughts about the source of knowledge in experience, in order to enjoy episodic memories.

It might be queried whether we could ever have evidence that young children employ imagistic representations. One kind of evidence which

---

[6] "Or at least in context determined by": this is to take account of experiential memories of events which were experienced as a participant, but are remembered in a way in which one has an observer's standpoint on oneself participating in those events. This is the distinction psychologists sometimes label as that between "field" and "observer" memories. See Nigro and Neisser 1983; D. Schacter 1996: 21–2.

[7] On the distinction between episodic memory and generic representations, including script-like representations, see McCormack and Hoerl (1999). Perner himself has also made the point very forcefully, in Perner (forthcoming).

could support the hypothesis that they do would be the results of experiments in the style of imagery research on the times taken to answer certain questions. Suppose a child without the concept of experience remembers being in front of a certain layout of objects, say, remembers being just inside the entrance of the zoo, with the penguin pool, the ice-cream stand, and the phone booth ahead of him at various different angles. We can ask the child questions analogous to some of those which imagery researchers ask to help establish that a subject is using imagery rather than 'purely propositional' representations. We can ask, for instance, 'Were the penguin pool, the ice-cream stand, and the phone booth all in a straight line?' where this was a question not raised at the time of the experience. We could accumulate evidence from speed of responses that these questions were being answered on the basis of imagery rather than semantic memory, or use of propositional list structures.

Kosslyn of course famously asked his subject to use imagery. In one experiment, for instance, he asked them "to imagine the map and mentally stare at a named location", and to imagine a moving black speck (Kosslyn 1980: 43). Obviously such techniques are not available in precisely the case of present concern, that in which the child does or may not have the concept of experiencing, imagining, or seeing. But the question "Were the pool, the ice-cream stand, and the phone booth in a straight line?" does not involve the concepts of experiencing, imagining, and seeing.

Even in the cases in which the instruction to imagine a moving black dot was important to Kosslyn, there may be analogues which do not involve exercise of the concept of experience on the part of the child, but which still involve the use of memory images to answer questions. We could ask the child: 'Starting from now, suppose you were walking from the entrance to the ice-cream stand, show me how long from now it would take you to get there, by pressing this button at the end of that time.' A linear function of time indicated to previously perceived distance would not of course be conclusive. Purely propositional representational of distance, with correct abilities to estimate how long it would take to walk them, would equally generate linearity. Evidence against these propositional hypotheses could, though, be found by applying in this case versions of the "intermediate state" and accompanying probe techniques familiar from other imagery studies, in particular those of Cooper and Shepard.[8] The child who is using experiential memory images in answering these questions will be able to answer correctly when probed with such further questions as 'What would you be level with now in walking to the ice-cream stand?'

To summarize the first problem: it seems to me that Perner and Ruffman's hypothesis attributes more than is required for a subject to have episodic memories.

---

[8] See Cooper and Shepard 1982; Cooper 1982.

Problem Two   As the first quotation from Perner and Ruffman shows, they envisage the distinctive feature of episodic memory as being the subject's possession of a representation whose content includes *I have seen/heard/ felt... it as being the case that p*. It seems to me that this proposal is open to a dilemma. Either the content *p* already contains some memory-demonstrative, or it does not. By a memory-demonstrative, I mean a demonstrative in thought made available by a memory image of the thing to which the demonstrative refers. I may represent it as being the case that I have seen that beach, with that yacht pulled up on the sand, with that headland in the distance. Here 'that beach', 'that yacht', 'the sand', 'that headland' all express memory demonstratives. Maybe it is plausible that, when I have the concept of experience, and have episodic memories, I am at least in a position to form such representations. But such representations about what I experienced could hardly be explanatory of the existence of the episodic memory. The memory demonstratives are available only because I have the episodic memory. They could not exist without it, any more than perceptual-demonstratives could be available without the perceptual experiences which make them available.

One might compare the proposal that the difference between perceptual experience and other informational states about one's environment is the subject's having a representation with such a content as *I am seeing it as being the case that that book is on that desk*. If the perceptual experience did not exist, the perceptual demonstratives 'that book', 'that desk' would not be available either.

Suppose, to give the other horn of the dilemma, that the embedded content *p* to which the operator 'I have experienced it as being the case that——' (or its like) is applied do not contain any memory demonstratives. This would avoid the problem just described, but at the price of giving something which is far from enough for episodic memory. I can know I have experienced it as being the case that I am in a certain part of Birmingham, without remembering being in that part of Birmingham. This is just more semantic memory: higher-order, certainly, but still semantic. Applying an operator to a content delivered by semantic memory will not take the subject beyond semantic memory.

In short: if the operator is applied to a content which involves memory demonstratives, then there must already be a kind of episodic memory whose existence is not to be explained in terms of the application of such operators; and if the content does not contain such operators, application of operators to it will never yield episodic memory from those materials. This second horn of the dilemma, if the argument here is sound, entails that Perner and Ruffman's own particular proposal for what is involved in episodic memory of seeing a word on a list is insufficient. In the first of the quotations above, the "additional fact" which they say must be represented is that one has seen the word "butter" on the list and that, therefore, one knows that the word has appeared on the list. One could represent all this to oneself, and indeed know it to be so, without remembering seeing the word 'butter' on the list. This could be the case, for instance, if one knew that one's present merely semantic memory (and knowledge) that it was on the list is a result of having seen it on the list, even though one cannot now remember seeing it on the list.[9]

In response to Problems One and Two, one can conceive of a reply on behalf of Perner and Ruffman's position. In response, they might say that their theory concerns only remembering in the class of cases of which it is definitional, or a priori constitutive, that remembering involves exercise of the concept of experience, and involves representing oneself as having experienced certain events or states of affairs earlier. This defence of their position would not dispute the case made above that there can be episodic memories which do not involve exercise of the concept of experience, or do not involve beliefs about its status as a source of knowledge. On this defence, Perner and Ruffman are simply concerned with a proper subset of episodic memories, namely those which do involve exercise of the concept of experience, and appreciation of its role in the production of knowledge. This defence, though, no longer explains all the phenomena of childhood amnesia. There is childhood amnesia not only in respect of the restricted subclass of memories just mentioned, but in respect of some wider class of episodic memories more generally, even when they do not involve exercise of the concept of experience, nor the exercise of any knowledge about the relations between experience and knowledge.

It is significant, and Perner and Ruffman are surely right to emphasize, that childhood amnesia begins to disappear as concepts of experience—seeing, hearing, and the rest—and understanding of their instances to a subject's own knowledge—begin to feature in the child's repertoire. This is not a matter on which a philosopher is in any position to offer new experimental evidence. But one can conceive of several other explanations consistent with the data they cite. One hypothesis worth investigating is that acquisition of the concept of experience, and a thinker's application of it to his own states, including applications within the scope of 'I remember——', facilitates remembering itself. Such facilitation could also explain the improvements in free-recall tasks, and in source-monitoring, amongst those children who have acquired concepts of experience.[10]

If Perner and Ruffman's Hypothesis were correct, then it would be impossible for someone without the concept of experience to gain noninferential

---

[9] It should be emphasized that Perner and Ruffman are not concerned only with some subpersonal representation of something as experience. The sort of evidence offered in support of children's possession of the concept is their ability to say, for instance, that they know something because they have seen it to be so. See Perner and Ruffman's discussion, 1995: 518.

[10] On the improvement in free-recall tasks, see Wimmer, Hogrefe, and Perner 1988. On the improvement in source-monitoring, see Lindsay, Johnson, and Kwon 1991.

knowledge about his own past, and that of the world, from his episodic memories. Yet that seems to be in principle possible. (It may also actually be the case for some children below the age of about three-and-a-half years.) A partially externalist account of what it is for an episodic memory to have a past-tense content does not require the rememberer to have the concept of experience and memory. Such an account will have a batch of requirements concerning the normal sensitivity of such memories to past states of affairs themselves, as things actually are, together with requirements about the rememberer's ability to integrate their contents into a history of himself and the world. None of these requirements demand the rememberer's possession of the concept of experience (though they certainly contain the seeds for its development). Empirical psychological theories which permit the full integration of the epistemology and the metaphysics of the past should not require such possession either.

Conclusion

I have perforce been brief on many matters which deserve a much more extended treatment. I hope, though, that I have done something to make it plausible that we cannot pursue our epistemology, our metaphysics, our theory of concepts, and our psychology without taking proper account of their interrelations. A satisfactory treatment must integrate all of them.

12 Connecting Vision with the World: Tracking the Missing Link

Zenon W. Pylyshyn

You might reasonably surmise from the title of this chapter that I will be discussing a theory of vision. After all, what is a theory of vision but a theory of how the world is connected to our visual representations? Theories of visual perception universally attempt to give an account of how a proximal stimulus (presumably a pattern impinging on the retina) can lead to a rich representation of a three-dimensional world and thence to either the recognition of known objects or to the coordination of actions with visual information. Such theories typically provide an effective (i.e. computable) mapping from a 2D pattern to a representation of a 3D scene, usually in the form of a symbol structure. But such a mapping, though undoubtedly the essential purpose of a theory of vision, leaves at least one serious problem that I intend to discuss here. It is this problem, rather than a theory of vision itself, that is the subject of this chapter.

The problem is that of connecting visual representations to the world in a certain critical way. This problem occurs for a number of reasons, but for our purposes I will emphasize just one such reason: the mapping from the world to our visual representation is not arrived at in one step, but rather it is built up incrementally. We know this both from empirical observations (e.g. percepts are generally built up by scanning attention and/or one's gaze) and also from theoretical analysis (e.g. Ullman 1984 has provided good arguments for believing that some relational properties, such as the property of being inside or on the same contour, have to be acquired serially by scanning a display). Now here is one problem that arises immediately. If the representation is built up incrementally, we need to know that a certain part of our current representation refers to a particular individual object in the world. The reason is quite simple. As we elaborate the representation by uncovering new properties of a scene that we have partially encoded we need to know where (i.e. to which part of the representation) to attach the new information. In other words we need to know when a certain token in the existing representation should be taken as corresponding to the *same individual object* as a particular token in the new representation, so that we can attribute newly noticed properties to the representation of the appropriate individual objects.

This chapter is based on work-in-progress being conducted jointly with Jerry Fodor. I wish to acknowledge his contribution, as well as his critical reading of an earlier draft. The errors in this chapter are not only mine, but are probably due to my refusing to accept his advice on certain points.

Take a concrete example. Suppose the representation of a scene takes the form of a conceptual structure whose content corresponds roughly to that of the English sentence fragment, '... four lines forming a parallelogram, with another similar parallelogram directly below it, arranged so that each vertex of the top parallelogram is connected by a straight line to the corresponding vertex of the parallelogram below it'. Although we can infer that there are twelve lines in this figure, we don't have a way to refer to them individually. We can't say which lines are referred to in the first part of the description ('... four lines forming a parallelogram'), which lines are the ones that connect the two parallelograms, and so on. Without identifying particular lines we could not add further information to elaborate the representation. If, for example, on further examination, we discover that certain of the lines are longer than others, some are colored differently, some vertices form different angles than others, and so on, we would need to connect this new information to representations of particular objects in the interim representation. Conjunctions of properties (e.g. red, right-angled, lower, etc.) are defined with respect to particular objects, so individual objects must be identified in order to determine whether there are property conjunctions. The question is, how can a representation identify particular objects?

Let's look at this example more closely. The content of the descriptive sentence in the above example might refer to the Necker cube shown on the left in Figure 12.1 (where the parallelograms in question are the figures EFGH and ABCD).

Now suppose that at some point you notice, as most people do sooner or later, that the face labeled FGBC is a square that appears to lie in front of (i.e.

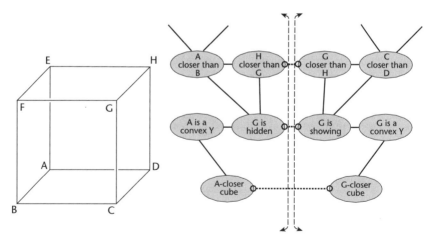

Figure 12.1 Different forms of descriptive representations of the figure on the left. The middle form, used by some connectionist systems, is based on Feldman and Ballard 1982.

is closer to the viewer than) the square EHAD. How would we add that information to a representation whose content is like that of the sentence quoted earlier? In order to add information to a representation we need to relate the information to representations of particular elements in the figure. That's why in this example, as in using diagrams in general, we label lines or vertices. Whatever form the visual representation takes it must allow the recovery of particular individual objects or parts referred to in that represent-ation much as though they were labeled. What constraints does that impose on a representation? Can a purely descriptive representation (i.e. a description with quantifiers but no names or singular terms) suffice? This is a question that gets into much deeper issues than ones I can address in any detail here. Yet it needs to be addressed at least briefly insofar as I will argue that visual representations need something like demonstratives or names in order to allow incremental elaboration (and for other reasons as well).

Common forms of representations of a simple figure such as a Necker cube are shown in Figure 12.1. In order to be able to augment the description over time it would be necessary to pick out particular token objects (lines or vertices) that appear in the representation. Assuming that we have not labeled every relevant point in the figure (after all, the world does not come con-veniently labeled), a possible way in which a purely descriptive representation could pick out individuals is by using definite descriptions. It could, for example, assert things like "the object x that has property P" where P uniquely picks out a particular object. In that case, in order to add new information, such as that this particular object also has property Q, one would add the new predicate Q and also introduce an identity assertion, thus asserting that $P(x) \wedge Q(y)$ and $x \equiv y$ (and, by the way, adding this new compound descriptor to memory so that the same object might be relocated in this way when a further new property of that object is later noticed).[1] But this is almost certainly not how the visual system adds information. This way of adding information would require adding a new predicate Q to the representation of an object that is *picked out by a certain descriptor*. To do that would require first recalling the description under which x was last encoded, and then conjoining to it the new descriptor and identity statement. Each new description added would require retrieving the description under which the object in question was last encoded.

---

[1] Strictly speaking the description that uniquely picks out a certain object at a particular time is a quantified expression of the form: $\exists x P(x)$, where P is the unique property of the object in question. When an additional predicate Q that pertains to the same object is to be added, the unique descriptor is retrieved and the new expression added: $\exists x \, \exists y \, \{P(x) \wedge Q(y) \wedge x = y\}$. If a further property R of the same object is detected at some later time, the last expression must be matched to the object at which R is discovered and its descriptor updated to the expression $\exists x \, \exists y \, \exists z \, \{P(x) \wedge Q(y) \wedge R(z) \wedge x = y \wedge y = z\}$. This continual updating of descriptors capable of uniquely picking out objects is clearly not a plausible mechanism for incrementally adding to a visual representation. It demands increasingly large storage and retrieval based on pattern matching.

The alternative to this unwieldy method is to allow the descriptive apparatus to make use of singular terms or names or demonstratives.[2] If we do that, then adding new information would amount to adding the predicate Q(a) to the representation of a particular object a, and so on for each newly noticed property of a. Empirical evidence for object-based attention (see the discussion in the last section of this chapter and, for example, Baylis and Driver 1993) suggests that the visual system's property detectors (e.g. Q-Detectors) recognize instances of the property Q *as a property of a particular visible object*, such as object a, so that this is the most natural way to view the introduction of new visual properties by the sensorium. In order to introduce new properties in that way, however, there would have to be a non-descriptive way of picking out a, such as a singular term or a name or a demonstrative. This is, in effect, what labeling objects in a diagram does through external means and what demonstrative terms like 'this' or 'that' do in natural language.[3] This alternative is *prima facie* the more plausible one since it is surely the case that when we detect a new property we detect it as applying to *that* object, rather than as applying to some object in virtue of its being the object with a certain (recalled) property.[4] Such intuitions, however, are notoriously unreliable so later in this chapter I will examine empirical evidence which suggests that this view is indeed more likely to be the correct one. For example, I will describe studies involving multiple-object tracking that make it very unlikely that objects are tracked by regularly updating a description that uniquely picks out the objects. In these studies the only unique descriptor available is location, and under

---

[2] Chris Peacocke has pointed out to me that calling this index-binding mechanism a name is misleading since names are used primarily to allow us to think about objects in their absence. The exact terminology that should be used in order to avoid misunderstanding is unclear. The term 'demonstrative' implies a natural language context and an intention on the part of a speaker to demonstrate that object, which is not the case in the mental index I have in mind. It seems that the term 'visual demonstrative' has been used with precisely the sense I have in mind, so I will henceforth confine myself to this terminology (along with the more technical phrase 'index binding' which invokes the theoretical mechanism I have proposed and discussed elsewhere).

[3] Notice that the need for demonstratives remains even if the representation were picture-like instead of symbolic, so long as it was not an exact and complete copy of the world but was built up incrementally. If the picture depicts some state of affairs in the world, we still have the problem of deciding when two pictorial bits are supposed to depict the same object. We still need to decide when two picture-fragments are supposed to depict the same object (even though they may look different) and when they are supposed to depict different objects. This is the same problem we faced in the case of symbolic representations. We don't know whether the thing in the picture that is depicted as having the property P is the thing to which we must now add the depiction of the newly noticed fact that it also has property Q. Without a solution to that puzzle we don't know to which part of the picture to add newly noticed properties.

[4] There is another alternative for picking out objects that I will not discuss here because the evidence I will cite suggests that it is not the correct option for visual representations. This alternative assumes the existence of demonstratives, as we have done, except the demonstratives in question are place demonstratives or locatives, such as 'this place'. Such an apparatus would allow the unique picking-out of objects based on their locations and would overcome the problem with the pure descriptivist story that I have been describing. That alternative is compatible with the view presented here although, as I will argue, the idea that object individuation is mediated by location alone does not seem to be supported by the empirical data.

certain plausible assumptions the evidence shows that it is very unlikely that the coordinates of the points being tracked are being regularly updated so that tracking is based on maintaining identity by updating descriptions.

There are a number of other reasons why a visual representation needs to be able to pick out individuals the way demonstratives do (i.e. independent of their particular properties). For example, among the properties that are extracted (and presumably encoded in some way) by the visual system are a variety of relational predicates, such as

$$\text{Collinear } (X_1, X_2, \ldots X_n) \text{ or Inside } (X_1, C_1) \text{ or Part-of}(F_1, F_2)$$

and so on. But these predicates apply over distinct individual objects in the scene independent of what properties these individuals have. So in order to recognize a relational property involving several objects we need to specify which objects are involved. For example, we cannot recognize the Collinear relation without somehow picking out which objects are collinear. If there are many objects in a scene only some of them may be collinear so we must associate the relation with the objects in question. This is quite general since properties are predicated of things, and relational properties (like the property of being "collinear") are predicated of several things. So there must be a way, independent of the process of deciding which property obtains, of specifying which objects (in our current question-begging sense) have that property. Ullman, as well as a large number of other investigators (Ballard *et al.* 1997; Watson and Humphreys 1997; Yantis 1998; Yantis and Johnson 1990; Yantis and Jones 1991) talk of the objects in question as being "tagged" (indeed, "tagging" is one of the basic operations in Ullman's theory of visual routines). The notion of a tag is an intuitive one since it suggests a way of *marking objects* for reference purposes. But the operation of tagging only makes sense if there is something on which a tag literally can be placed. It does no good to tag an internal representation since the relation we wish to encode holds in the world and may not hold in the representation. But how do we tag parts of the world? What we need is what labels gave us in the previous example: a way to name or refer to individual parts of a scene *independent of their properties or their locations*.

What this means is that the representation of a visual scene must contain something more than descriptive (or pictorial—see n. 3) information in order to allow re-identification of particular individual visual elements. It must provide what natural language provides when it uses names (or labels) that uniquely pick out particular individuals, or when it embraces demonstrative terms like 'this' or 'that'. Such terms are used to indicate particular individuals. This assumes that we have a way to *individuate[5] and keep track of*

---

[5] As with a number of terms used in the context of early vision (such as the term 'object'), the notion of individuating has a narrower meaning here than in the more general context where it refers

particular individuals in a scene even when the individuals change their properties, including their locations. Thus what we need are two functions that are central to our concern here: (*a*) we need to be able to pick out or individuate distinct individuals (following current practice, we will call these individuals *objects*) and (*b*) we need to be able to refer to these objects as though they had names or labels. Both these purposes are served by a primitive visual mechanism that I call a *visual index*. So what remains is for me to provide an empirical basis for the claim that the visual system embodies a primitive mechanism of the sort I call a *visual index*.

Individuating and Tracking Primitive Visible Objects: Multiple Object Tracking Studies

Perhaps the clearest way to see what I mean, when I claim that there is a primitive mechanism in early vision that picks out and maintains the identity of visible objects, is to consider a set of experiments, carried out in my laboratory, to which the ideas of visual individuation and identity maintenance were applied. The task is called the *Multiple Object Tracking (MOT) Task* and is illustrated in Figure 12.2.

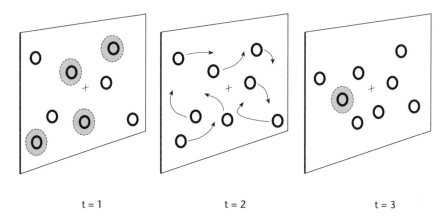

t = 1   t = 2   t = 3

Figure 12.2 Illustration of a typical multiple-object-tracking (MOT) experiment. In the first panel a subset of identical elements are flashed to identify them as the "targets" to be tracked. In the middle panel the elements move around at random (usually without colliding), and in the third panel one of the elements is selected (probed). The subjects' task is to say whether that element was one of the targets.

---

not only to separating a part of the visual world from the rest of the clutter (which is what is meant by individuate here), but also providing identity criteria for recognition instances of that individual. As is the case with objecthood and other such notions, I am here referring primarily to primitive cases—i.e. ones provided directly by mechanisms in the early vision system (in the sense of Pylyshyn 1999) and not constructed from other perceptual functions.

In a typical experiment, subjects are shown a screen containing anywhere from twelve to twenty-four simple identical objects (points, plus signs, figure-eight shapes) which move across the entire visual field in unpredictable ways without colliding. A subset of these objects is briefly rendered distinct (usually by flashing them on and off a few times). The subject's task is to keep track of this subset of objects (called "targets"). At some later time in the experiment (say ten seconds into the tracking trial) one of the objects is again flashed on and off. The subject must then indicate whether or not the flashed (probe) figure was one of the targets. A large number of experiments, beginning with studies by Pylyshyn and Storm (1988) have shown clearly that subjects can indeed track up to five independently moving identical objects. Moreover, we were able to argue that the motion and dispersion parameters of the original Pylyshyn and Storm experiment were such that tracking could not have been accomplished using a serial strategy in which attention is scanned to each figure in turn, storing its location, and returning to find the figure closest to that location on the next iteration, and so on. Based on some weak assumptions about how fast focal attention might be scanned and based on actual data on how fast the objects actually moved and how close together they had been in this study, we were able to conclude that such a serial tracking process would very frequently end up switching to the wrong objects in the course of its tracking. This means that the moving objects *could not have been tracked using a unique stored description of each figure,* inasmuch as the only possible descriptor that was unique to each figure at any particular instant in time was its location. If we are correct in arguing from the nature of the tracking parameters that stored locations cannot be used as the basis for tracking then all that is left is the figure's identity or *individuality.* This is exactly what I claim is going on—tracking by maintenance of a primitive perceptual individuality.

Recently a large number of additional studies in our laboratory (Pylyshyn 1998; Sears and Pylyshyn 2000, McKeever 1991; Scholl and Pylyshyn 1999) and elsewhere (Intriligator and Cavanagh 1992; Yantis 1992; and others) have replicated these multiple object tracking results using a variety of different methods, confirming that subjects can successfully track around four or five independently moving objects. The results also showed that merely widening one's breadth of attention (as assumed in the so-called zoom-lens model of attention spreading : Eriksen and St James 1986) would not account for the data. Performance in detecting changes to elements located inside the convex hull outline of the set of targets was no better than performance on elements outside this region, contrary to what would be expected if the area of attention were simply widened or shaped to conform to an appropriate outline (Pylyshyn *et al.* 1994). Using a different tracking methodology, Intriligator and Cavanagh (1992) also failed to find any evidence of a "spread of attention" to regions between targets. It appears, then, that items can be tracked despite the lack of distinctive properties (and indeed when their properties are changing) and despite constantly changing locations and unpredictable motions. Taken together these studies imply a notion of primitive visible object as a category induced by the early visual system, preceding the recognition of properties and preceding the evaluation of any visual predicate.

The multiple object tracking task exemplifies what I mean by "tracking" and by "maintaining the identity" of objects. It also operationalizes the notion of "primitive visible object"—a primitive visible object is whatever attracts a visual index and allows multiple object tracking (visual indexes have also been called FINSTs, which once stood for FINgers of INSTantiation because they bind or instantiate variables, acting like pointing fingers). Note that this is a highly mind-dependent definition of objecthood. Objecthood and object-identity are defined in terms of a causal perceptual mechanism. A certain sequence of object-locations will count as the movement of a single object if the early (pre-attentive) visual system groups it this way—that is, if it is so perceived—whether or not we can find a physical property that is invariant over this sequence and whether or not there exists a psychologically plausible description that covers this sequence. The visual system may also count as one individual object certain kinds of disappearance and reappearance of visual objects. For example, Yantis has shown that when an object disappears either for a very short time *or* under conditions where it is seen to have been occluded by an opaque surface, the visual system treats the object as though it continued to exist. Similarly, Scholl and Pylyshyn (1999) have shown that, if the objects being tracked in the MOT paradigm disappear and reappear in certain ways, they are tracked as though they had a continuous existence and a smooth trajectory. If they disappear and reappear by deletion and accretion along a fixed contour, the way they would if they were moving behind an occluding surface (even if the edges of the occluder are not invisible), then they are tracked as though they were continuously moving objects. Performance in the MOT task does not deteriorate if targets disappear in this fashion although it suffers dramatically if targets suddenly go out of existence and reappear, or if they slowly shrink away and then reappear by slowly growing again at exactly the same place as they had accreted in the occlusion condition.

A Theory of Visual Indexing and Binding: The FINST Mechanism

I now briefly review the theory of the Indexing mechanism for I intend it to serve a major function—that of providing the missing link alluded to in my title. The basic motivation for postulating indexes is that, as we saw at the beginning of this essay, there are a number of reasons for thinking that individual objects in the field of view must first be *picked out* from the rest of the visual field and the identity of these objects *qua individuals* must be maintained or tracked despite changes in the individual's properties, including its location in the visual field. Our proposal claims that this is done *primitively* without identifying the object through a unique descriptor. The object in question must be segregated from the background or picked out as an individual (the Gestalt notion of making a figure-ground distinction is closely related to this sort of "picking out"). Until some piece of the visual field is segregated and picked out, no visual operation can be applied to it since it does not exist as something distinct from the entire field.

In its usual sense (at least in philosophy), picking out an individual requires having criteria of individuation—that is, requires having a sortal concept. How can we track something without re-recognizing it as the same thing at distinct periods of time, and how can we do that unless we have a description of it? My claim is that just as the separation of figure from ground (the "picking-out") is a primitive function of the architecture of the visual system, so also is this special sort of pre-attentive tracking. What I am proposing is not a full-blooded sense of identity-maintenance, but a sense that is relativized to the basic character of the early visual system. The visual system cannot in general re-recognize objects as being the same without some descriptive apparatus, but it can track in a more primitive sense, providing certain conditions are met (these conditions include continuity of motion or else the presence of local occlusion cues such as those mentioned above in discussing the Yantis and the Pylyshyn and Scholl results).

What this means is that our theory is concerned with a sense of *picking out* and *tracking* that are not based on top–down *conceptual* descriptions, but are given pre-conceptually by the early visual system, and in particular by the FINST indexing mechanism. Moreover, the visual system treats the object so picked out as distinct from other individuals, independent of what properties this object might have or whether the properties are changing in unpredictable ways. If two different objects are individuated in this way they remain distinct as far as the visual system is concerned. Moreover, they remain distinct despite certain changes in their properties, particularly changes in their location. Yet the visual system need not know (i.e. need not have detected or encoded) any of their properties in order to implicitly treat them as though they were distinct and enduring visual tokens. Of course there doubtless are properties, such as being in different locations or moving in different ways or flashing on and off that allow indexes to be assigned to these primitive objects in the first place. But none of these properties define the objects—they are not *essential properties*. What is an essential property is that, given the structure of the early visual system, the object attracted and maintained an index. My claim is that to index *x*, *in this primitive sensory sense*, there need not be any concept, description, or sortal that picks out *x*'s by type.[6] The individuals picked out in this way by the early visual system (by

---

[6] The claim is that there is a mechanism in the early (pre-conceptual) visual system that latches onto certain entities (perhaps I should say "events") for purely causal reasons, not because those entities meet conditions provided by a cognitive predicate—i.e. not because they constitute

a mechanism that I will describe below) are what I am referring to here as *primitive visible objects*. I use this technical terminology to distinguish these primitive visible objects from the more general sense of object, which might include invisible things, abstract things (like ideas), and other more usual notions of object, such as tables and chairs and people—which writers like Hirsch (1982) and Wiggins (1980) and others have argued, *does* require sortal concepts to establish criteria of identity. My concern here will be with objects that are in the first instance defined in terms of the individuation (or clustering) and indexing mechanism of the early visual system, although this sort of individuation, I claim, must form the basis for full-fledged individuation. The latter cannot be conceptual "all the way down" on pain of infinite regress. My claim, then, is that certain mechanisms of the early visual system lead to the automatic individuation of a small number of primitive visible objects and to the tracking of such individuals over certain sorts of change of time and space.

The basic idea of the FINST indexing and binding mechanism is illustrated in Figure 12.3. A series of proximal causes leads from certain kinds of visible events, via primitive mechanisms of the early visual system, to certain conceptual structures (which we may think of as symbol structures in long-term memory). This provides a mechanism of reference between a visual representation and what we have called primitive visible objects in the world. The important thing here is that the inward arrows are purely causal and are instantiated by the non-conceptual apparatus of what I have called *early vision* (Pylyshyn 1999). Under certain conditions this mechanism results in a link that exhibits a certain continuity or persistence, thus resulting in its counting as the *same link*. It is tempting to say that what makes it continuous is that it keeps pointing to the *same thing*, but according to our view this is circular since the only thing that makes it the same thing is the very fact that the index references it. There is no other sense of "sameness" so that "primitive visible object" as we have defined it is thoroughly mind-dependent.

By virtue of this causal connection, the conceptual system can *refer to* any of a small number of primitive visible objects. It can, for example, interrogate them to determine some of their properties, it can evaluate visual predicates (such as Collinear) over them, it can move focal attention to them, and so on. The function that I am describing is extremely simple and only seems complicated because ordinary language fails to respect certain distinctions (such as the distinction between individuating and recognizing, indexing and

---

instances of a certain concept. In other words, if P(x) is a primitive visual predicate of x, then the x is assumed to have been independently and causally bound to a "primitive visible object". Although this sort of latching or seizing by primitive visible objects is essentially a bottom–up process, this is not to say that it could not in some cases (perhaps in most cases) be guided by intentional processes, such as perhaps scanning one's attention until a latching event is located or an object meeting a certain description is found. For example, it is widely assumed (Posner, Snyder, and Davidson 1980) that people can scan their attention along some path (by simply moving it continuously through space like a spotlight beam) and thereby locate certain sorts of object. A possible consequence of such scanning is that an index may get assigned to some primitive objects encountered along the way.

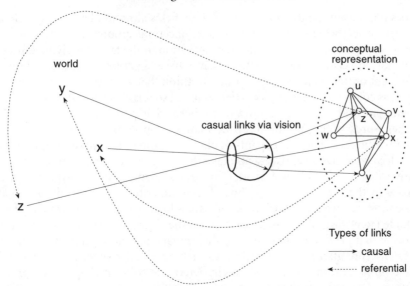

Figure 12.3 Visual Indexes (FINSTs) are drawn to primitive visual objects by a data-driven causal chain that passes through their proximal (retinal) projection. Once assigned, these FINST indexes can be used to refer to the distal object in question. They link elements of the conceptual representation (shown here as a network) with primitive objects in the world in a direct (unmediated) manner, not by virtue of how the object is conceptualized.

knowing where something is, and so on). Elsewhere (Pylyshyn 1999) I provide an extremely simple network, based on the Koch and Ullman (1984) winner-take-all neural net, which implements such a function.

Linking Visual Representations to Primitive Visual Objects

What we have described is a mechanism for picking out, tracking, and providing cognitive access to what we call an object (or, more precisely, a *primitive visible object*). The notion of an *object* is ubiquitous in cognitive science, not only in vision but much more widely. I might also note that it has been a central focus in developmental psychology where people like Susan Carey (this volume, Chapter 3), Fei Xu (1997), Alan Leslie (Leslie *et al.* 1998) have studied "a child's concept of object". Similarly, many studies have shown that attention is allocated primarily by individual visual object, rather than in terms of regions (Baylis and Driver 1993), a finding that is also supported by evidence from clinical neuroscience, where it has been argued that deficits such as unilateral neglect must be understood as a deficit of object-based attention rather than space-based attention (Driver and Halligan 1991). Space does not permit me to go into any of these fields although I am engaged in a larger project where I do examine the connections among these uses of the term 'object'. But I would like to draw attention to the fact that giving objects the sort of central role in vision that I have described suggests a rather different ontology. Just as it is natural to think that we apprehend properties such as color and shape as *properties of objects*, so it is also natural to think that we apprehend objects as a kind of property that particular *places* have. In other words we usually think of the matrix of space-time as being primary and of objects as being occupants of places and times. Everyone from Kant to modern cognitive scientists tacitly take this for granted—that's (in part) why it is so natural to think of mental images as having to be embedded in real space in the brain. Yet the findings I have described in the study of visual attention (as well as other areas of psychological research to which I will allude later) suggests an alternative and rather intriguing possibility. It is the notion that *primitive visible object* is the primary and more primitive category of early (pre-attentive) perception, so that we perceive objecthood first and determine location the way we might determine color or shape—as a property associated with objects.[7] If this is true then it raises some interesting possibilities concerning the nature of the mechanisms of early vision. In particular it adds further credence to what I argued is needed for independent reasons—some way of referring directly to primitive visible objects without using a unique description under which that object falls. This is the mechanism I referred to as a visual index or a visual demonstrative (or FINST).

Notice that when I am careful I hedge my use of the term *object* in making this claim, as I must because what I have been describing is not the notion of an object in the usual sense of an individual. An individual, as you all know, is a sortal concept whose individuation depends on assuming certain conceptual categories. But our notion does not assume the use of any concepts. The individuals that are picked out by the visual system and tracked primitively are something less than full-blooded individuals. Yet because they are what our visual system gives us through a brute causal mechanism—because that is its nature—it serves as the basis for all real individuation. As philosophers like

---

[7] The idea that location is encoded like any other property and is not used to uniquely pick out objects is controversial. For example, it is widely held that location is special and used to pick out objects. There have been a number of studies (reviewed in Pashler 1998) showing that in those cases when an object is correctly identified its location generally can be correctly reported. However, what these studies actually show is that for objects whose shapes (or in some cases color) can be correctly reported, their location can usually also be reported. From our perspective this only shows that there is a precedence ranking among the various properties of an object that are recorded and reported and that rough location may be higher on the ranking than other properties. What the experiments do not show (contrary to some claims) is that in order to detect the presence of an object one must first detect its location. The studies described herein (dealing with multiple Indexing) suggest ways to decide whether an object has been detected in the relevant sense (i.e. individuated and indexed, though not necessarily recognized). The theoretical position sketched here entails that one can index an object without encoding its location. There are, so far as I know, no data one way or another regarding this prediction.

Wiggins (1980) and Hirsch (1982) have argued, you cannot individuate objects in the full-blooded sense without a conceptual apparatus—without sortal concepts. But similarly you cannot individuate them with *only* a conceptual apparatus. Sooner or later concepts must be grounded in a primitive causal connection between thoughts and things. The project of grounding concepts in sense data has not fared well and has been abandoned in cognitive science. However, the principle of grounding concepts in perception remains an essential operation if we are not to succumb to an infinite regress. Visual indexes provide a putative grounding for basic objects and we should be grateful because without them (or at any rate something like them) we would be lost in thought without any grounding in causal connections with the real-world objects of our thoughts. With indexes we can think about things (I am sometimes tempted to call them FINGs since they are interdefined with FINSTs) without having any concepts of them: one might say that we can have *demonstrative thoughts*. And nobody ought to be surprised by this since we know that we can do this: I can think of this thing here without *any description* under which it falls. And, perhaps even more important, because I can do that, I can reach for it.

Needless to say there are some details to be worked out so this is a work-in-progress. But I hope I have at least demonstrated that there is a real problem to be solved in connecting visual representations to the world and that, whatever the eventual solution turns out to be, it will have to respect a collection of facts some of which I have sketched in this chapter. Moreover any visual or attentional mechanism that might be hypothesized for this purpose will have far-reaching implications, not only for theories of situated vision, but also for grounding the content of visual representations and perhaps for grounding perceptual concepts in general.

13 Rationality and Action

John R. Searle

1. Introduction

A certain conception of rationality and of the relationship between rationality
and decision making is pervasive in our philosophical tradition. This Classical
Model, as I will call it, begins with Aristotle's claim that deliberation is always
about means and never about ends; it receives its most famous formulation in
Hume's slogan: Reason is and ought to be the slave of the passions; and it has
its most sophisticated contemporary expression in mathematical decision
theory. I believe that the Classical Model is mistaken, that at best it describes
only a very special class of cases and does not give us a general account of the
role of rationality in action. I do not have the space here to expose all of its
weaknesses but will concentrate on one major feature.

It is an essential to the model that all rational bases for action are dependent
on preexisting desires of the agent. According to the model actions are caused
by beliefs and desires and beliefs and desires are the only possible reasons for an
action. Facts in the world as such can never be reasons for action. Reasons can
only be beliefs about those facts or desires regarding those facts. So, for
example, the fact that someone is starving is not a reason for doing anything,
only the fact that one believes that he is starving, and that, for example, one
desires to feed those who are starving, would lead one to do anything about it.
Rationality is always a matter of deliberating regarding our beliefs and desires
and the gap between the desire and the action have to be filled by the belief. On
this conception there cannot be any reasons independent of our desires. All
reasons for action are desire-dependent. All rationality in action is a matter of
deliberating on how to best satisfy our desires. Desires here have to be
construed broadly, to include not only wishes, but various moral evaluations,
principles, preferences, dispositions, and inclinations that one might have.

By way of beginning an attack on the classical model, I want to show how
we create desire-independent reasons for action. How can free rational agents
create desire-independent reasons for action? I believe that the answer to this
question is as simple as the question itself: free rational agents can create
desire-independent reasons by acting with the intention that their action
should create such reasons. The intentional performance of an act, with the
intention that that performance should create a reason, is, in certain contexts,
sufficient to create a reason. The interest of the investigation lies in showing
how the process works in detail, why it is important, and what bearing it has
on practical rationality in general.

But before turning to practical reason, I want to note that the creation of desire-independent reasons is completely pervasive in our lives, especially our social lives, and extends far beyond the realm of practical reason as traditionally construed. Pretty much whenever we perform a speech act, such as making an apology, or a statement, or issuing a command, or a promise, we are creating some sorts of desire-independent reasons. The general feature of all these cases is that there is a class of actions such that the performance of the action with the intention that that performance should create a reason is sufficient for the creation of the reason. We are blinded to the pervasiveness of this feature by the fact that we are seldom just creating a reason, but typically we are creating a reason as part of doing something else, such as issuing an invitation to a party, selling a used car, or ordering a beer. In the following discussion I will try to show how to factor out the common element of creating a reason, and show how it relates to the special features of practical reason.

Consider a type of case which is fairly simple and does not involve practical reason. Whenever I make a statement I have a reason to speak truthfully. Why? Because a statement simply *is* a commitment to the truth of the expressed proposition. There is no gap at all between making a statement and committing oneself to its truth. That is, there are not two independent features of the speech act, first the making of a statement and second committing myself to its truth; there is only making the statement which is *eo ipso* a commitment to truth. Suppose you ask me, "What's the weather like outside?" And I say, "It's raining". I have *thereby* committed myself to the truth of the proposition that it is raining. The commitment to truth is perhaps most obvious in cases where the speaker is lying. If I don't in fact believe that it is raining, but I lie and say, "It's raining", my utterance is only intelligible to myself as a lie precisely because I understand that the utterance commits me to the truth of a proposition I do not believe to be true. And the lie can only succeed as a lie precisely because you take me to be making a statement and therefore committing myself to the truth of the expressed proposition. A similar point can be made about mistakes. Suppose I am not lying but am genuinely mistaken. I sincerely said it is raining, but all the same it is not raining. In such a case there still is something wrong with my speech act, namely it is false. But why is that wrong? After all, for every true proposition there is a false one. It is wrong, because the aim of a statment is to be true, and when I make a statement I commit myself to its truth.

There is no way that the Classical Model can account for these simple facts. The Classical Model is forced to say that there are two separate phenomena, the institution of statement making and then external to that is the principle that one should try to speak the truth. What reason have I to try to tell the truth when making a statement? The classical theorist is forced to say that I have *no reason at all just in virtue of making a statement*. The only reason I could have would be that I felt there would be bad consequences if I lied, or that I hold a moral principle, which is logically independent of the existence of the institution, to the effect that falsehood is wrong, or that I just felt an inclination to tell the truth, or had some other reason external to the institution. But on the Classical Model all such reasons are independent of the nature of statement making as such. I am claiming, on the contrary, that there is no way to explain what a statement is without explaining that the commitment to truth is *internal* to statement making.

Now I want to apply some of these lessons to practical reason. In many cases of practical reason, one creates a reason now for performing an act in the future. I believe the only way to understand how voluntary rational action can create reasons for future actions is to look at the matter from close up. So, let us consider the sorts of cases that happen in everyday life. Suppose I go into a bar and order a beer. Suppose I drink the beer and the time comes to pay for the beer. Now the question is, granted the sheer fact that I intended my behavior to place me under an obligation to pay for the beer, must I also have a reason independent of this fact such as a desire to pay for the beer, or some other appropriate element of my motivational set, in order to have a reason to pay for the beer? That is, in order to know if I have a reason to pay for the beer, do I first have to scrutinize my motivational set to see if there is any desire to pay for the beer, or to see if I hold any general principles about paying for beer that I have drunk? It seems to me the answer is, I do not. In such a case by ordering the beer, and drinking it when brought, I have already intentionally created a commitment or obligation to pay for it, and such commitments and obligations are species of reasons.

It is an absurdity of the Classical Model that it cannot account for such an obvious case. As in the case of truth telling, the defender of the Classical Model is forced to say that I only have a reason to pay for the beer if I can locate the relevant desire in my "motivational set". In opposition to this I want to claim that in this situation I have simply created a reason for myself to pay for the beer by ordering the beer and drinking it.

What exactly are the formal features of the situation that have enabled me to do that? What exactly are the truth conditions of the claim: Agent A has a desire-independent reason to perform act X in the future? What fact about him makes it the case that he has such a reason? Well, one sort of fact that would be sufficient is: Agent A has *created* a desire-independent reason for himself to perform act X in the future. So our question now boils down to, How does one go about such a creation? To answer the question, we have to ask, how did it seem to Agent A when he ordered the beer? Well, if I am the agent, the way that it seems to me is this: I am now performing an act which is such that I am in that very act trying to get the man to bring me a beer on the understanding that I am under an obligation to pay for it if he brings it. But if that is the intention then, by this very performance, if the man brings the beer, I have made it the case that I now have an obligation, and therefore a reason, which will be a reason for me to act in the future, and that reason that I now create will be independent of my other future desires. In such a case, a sufficient condition for an act to create a reason for me is that I intend that it create a reason for me.

The presupposition of the freedom of the agent is crucial to the case as I have described it. From the first-person point of view, by freely undertaking to create a reason for myself, I have already manifested a desire that such and such be a reason for me. I have already bound my will in the future through the free exercise of my will in the present. In the end all these questions must have trivial answers. Why is it a reason? Because I created it as a reason. Why is it a reason *for me*? Because I have freely created it as a reason for me.

But the skeptic will surely say at this point, "But so far you have given us no reason that any such creation of a reason should subsequently motivate an action. What is it about my act of creating a reason which is such that that act of creation together with rationality can motivate a subsequent action?" "How can desire-independent reasons motivate?" At this point we can say, the recognition that I have a desire-independent reason to pay for the beer can rationally motivate a desire to pay for it. The skeptical question, asked from the standpoint of the Classical Model, presupposes that only desires can be the rational grounds of reasons, but in such cases as I have described *the reason can be the ground of the desire and not conversely.* In ordinary English the correct description of this case is, "I want to pay for it because I have an obligation to pay for it." And the connection between reason, rationality, and desire is described as follows: the recognition of something as reason for an action is already the recognition of something as a reason for desiring to perform the action. But if I have voluntarily created something as a reason for me, then I have already recognized it as a reason for me.

2. Promising as a Special Case

Discussions of these issues usually spend a lot of time on promising, but I am trying to emphasize here that the phenomenon of agent-created desire-independent reasons is absolutely pervasive. You could not begin to understand social life without it, and promising is only a special and pure kind of case. However, the history of the debates about promising is revealing and I will be able to better explain what I am arguing for if I explain the obligation to keep a promise and expose some of the standard mistakes. The question is: What reason do we have for keeping a promise? And to the obvious answer: Promises are by definition creations of obligations; and obligations are by definition reasons for action, there is a follow up question: What is the source of the obligation to keep a promise?

There is no way that the Classical Model can account for the fact that the obligation to keep a promise is internal to the act of promising, just as the commitment to truth telling is internal to the act of statement making. That is, promising is by definition undertaking an obligation to do something. The tradition is forced to deny this fact, but in order to deny it the defenders of the Classical Model are typically forced to say some strange and, I believe, mistaken things. In this section I offer a brief list of the most common mistakes I have encountered.

There are three common but, I believe, mistaken claims that can be disposed of quickly. The first is to suppose there is some special *moral* obligation to keep a promise. On the contrary, if you think about it you will see that there is no special connection between promising and morality, strictly construed. If I promise to come to your party, for example, that is a social obligation. Whether it is a moral obligation as well would depend on the nature of the case, but for most parties I go to it would not be a moral obligation. Often we make promises where some grave moral issue is concerned but there is nothing about promising as such that guarantees that any promise at all involves moral issues. There is nothing in the practice of promising as such that guarantees that every obligation to keep a promise will be grave enough to be considered a moral obligation. One may make promises over matters that are morally trivial.

A second related mistake is to suppose that if you promise to do something evil there is no obligation at all to keep the promise. But this is obviously mistaken. The correct way to describe such cases is to say that you do indeed have an obligation to keep the promise but it is overriden by the evil nature of the promised act. This point can be proved by the method of agreement and difference: there is a difference between the person who has promised to do the act and the person who has not. The person who has made the promise has a reason that the person who has not made a promise does not have.

A third, and, I believe, the worst of the three mistakes, is to suppose that the obligation to keep a promise is only a *"prima facie"* obligation, as opposed to a flat outright obligation. This view was designed (by Sir David Ross) to try to evade the fact that obligations typically conflict and you often can't fulfill them all. When obligation A overrides obligation B, says Ross, B is only a *prima facie* obligation, not an honest-to-john outright, unconditional obligation. I have argued in detail elsewhere (Searle 1980*b*) that this view is confused and I won't repeat the arguments here except to say that, when B is overriden by some more important obligation, this does not show that B was not an all out, unconditional, etc. obligation. You can't override it if there there is nothing really there to override in the first place. *"Prima facie"* is an epistemic sentence modifier, not a predicate of obligation types, and could not possibly be an appropriate term for describing the phenomenon of conflicting obligations, where one is overriden by another. The theory of "prima facie obligations" is worse than bad philosophy, it is bad grammar.

I believe the following are the most common serious mistakes about the obligation to keep a promise, and they all derive in their different ways from an acceptance of the classical model:

Mistake 1: *The obligation to keep a promise is prudential. The reason for keeping this promise is that if I don't I will not be trusted in the future when I make promises.*

Famously, Hume held this view. But it is subject to a decisive, and equally famous, objection: On this account, in cases where no one knows of my failure to keep the promise, I would be under no obligation at all to keep it. The deathbed promise, made by the son in private to his dying father, on this view would involve no obligation at all because the son need not tell anyone about the promise.

Mistake 2: *The obligation to keep a promise derives from the acceptance of a moral principle to the effect that one ought to keep one's promises. Without such an acceptance the agent has no reasons, except perhaps prudential reasons, to keep a promise.*

The mistake here is the same as the mistake we found in the case of the commitment to truth when making a statement. The Classical Model tries to make the obligation in promising external to the act of promising, but then it becomes impossible to explain what a promise is, just as it becomes impossible to explain what a statement is if one tries to make the relation between stating and committing oneself to the truth as purely external. That is, the decisive answer to this objection is to point out that the relations between promising and obligations are internal. By definition a promise is an act of undertaking an obligation. It is impossible to explain what a promise is except in terms of undertaking an obligation.

Just as we saw in the case of statement making that the commitment to truth is most obviously revealed in the case of the person who deliberately lies, so in the case of promising we can see that the obligation is internal to the act of promising most obviously in the case of the person who makes an insincere promise. Suppose I make an insincere promise, a promise I make with no intention to keep. In such a case my act of deception is only fully intelligible to myself as a dishonest act, precisely because I understand that when I made the promise I was binding myself, undertaking an obligation to do the thing I promised to do. When I make a promise I am not hazarding a guess about what is going to happen in the future, I am rather binding my will as to what I am going to do in the future. My dishonest promise is intelligible to myself only as a promise in which I did undertake an obligation without any intention to fulfill the obligation I have undertaken.

Mistake 3 (this is a more sophisticated variant of mistake 2): *If obligations really were internal to promising, then the obligation to keep a promise would have to derive from the institution of promising. The fact that someone made a promise is an institutional fact, and any obligation would have to derive from the* *institution. But then what is to prevent any institution from having the same status? Slavery is as much an institution as promising. So if the view that promises create desire-independent reasons were right, then the slave would have as much an obligation as does the promissor, which is absurd. As institutions, promising and slavery are on all fours; the only difference as far as our present debate is concerned is that we happen to think the one is good, the other bad.*

This objection encapsulates the view of the Classical Model on this issue. The simplest answer to it is this: *The obligation to keep a promise does not derive from the institution of promising.* When I make a promise, the institution of promising is just the *vehicle*, the tool that I use to create a reason. The obligation to keep a promise derives from the fact that in promising I freely and voluntarily create a reason for myself. The free exercise of the will can bind the will, and that is a logical point that has nothing to do with "institutions" or "moral attitudes" or "evaluative utterances". This is why the slave does not have any reason to obey the slave owner, except prudential reasons. He has not exercised any freedom that has bound his will. Viewed externally the slave may look just like the contract laborer. They might even be given the same rewards. But internally it is quite different. The contract laborer has created a reason for himself which the slave has not created. To think that the obligation of promising derives from the institution of promising is as mistaken as to think that the obligations I undertake when I speak English must derive from the institution of English: unless I think English is somehow a good thing, I am under no obligations when I speak it.

3. A General Theory of Desire-Independent Reasons in Practical Reason

So far I have just tried to discuss a few cases and suggest some general conclusions. With this material in hand we can now attempt to state a more general theory. I want among other things to explain why the creation of desire-independent reasons by free rational agents in possession of a language and operating within institutional structures is so pervasive. This is what happens when you get married, order a beer in a bar, buy a house, enroll in a college course, or make an appointment with your dentist. In such cases you now invoke an institutional structure in such a way that you now create a reason for yourself to do something in the future regardless of whether you independently have a desire in the future to do that thing. And in such cases it is reason for you because you have voluntarily created it as a reason for you.

A general account of the role of reasons in practical rationality involves understanding at least the following five features: (1) freedom, (2) temporality, (3) the first-person point of view, (4) language and other institutional structures, and (5) rationality.

The last two of these are not peculiar to practical reason, but the first three have a special role in practical reason. Rationality in action, that is, practical reason, has at least these three features which make it quite unlike theoretical reason. First, it is essentially about human freedom; second, it is essentially about the ordering of time; and third, it is essentially from the point of view of the first person, in the sense that something can only be a reason for an action from the point of view of some agent. These are the major formal differences between theoretical reason and practical reason. Let us consider each in order:

Freedom

Rationality and the presupposition of freedom are coextensive. They are not the same thing, but actions are rationally assessable iff the actions are free. The reason for the connection is this: *Rationality must be able to make a difference.* If the act is completely determined, then rationality can make no difference. The person whose act is really caused by beliefs and desires, à la the Classical Model, is just acting as a compulsive and acts outside the scope of rationality altogether. But the person who freely *acts on* those same beliefs and desires, who freely makes them into *effective* reasons, acts within the realm of rationality. Freedom of action and the applicability of rationality are coextensive.

Furthermore, if the act is free, I can now have a reason to create and freely can create a reason such that the reason is a reason for me to do something in the future, regardless of whether I feel like doing it when the time comes. The ability to bind the will now can create a reason for the future act only because it is a manifestation of freedom.

Time

Theoretical reason statements are untensed in a way that practical reason statements are inherently tensed. "I am going to do act A because I want to make it the case that B" is essentially future referring, in the way that "Hypothesis H is substantiated by evidence E" is not essentially tensed at all. It is timeless, although of course, in particular instances, it may make reference to particular historical situations. For animals, there really are only immediate reasons, because without language you cannot order time.

The first-person point of view

The first-person point of view is essential to understanding practical reason in a way that it is not essential to understanding theoretical reason because actions are always performed by agents, and in order for an agent to behave rationally in the performance of an action a reason has to be a reason *for him*. And this is, in important respects, different from a reason for believing, because a reason for believing, if it is a justificatory reason, is a reason *for anybody.* A theory of practical reason has to explain how rationality can actually cause human bodies to move, and that can only be explained from the first-person point of view.

Language and other institutional structures

In order to create desire-independent reasons an agent has to have a language. There are several reasons for this but perhaps the most obvious ones are that without a language it is impossible to represent deontic relations and to order time in the required way. One has to have a way of representing the fact that one's present action creates a reason for a future action, and without linguistic devices it is impossible to represent the temporal and deontic relations in question.

In addition to language narrowly construed, that is, in addition to such speech acts as statement making or promising, there are extralinguistic institutional structures which also function in the creation of desire-independent reasons. So, for example, only if a society has the institution of property can there be desire-independent reasons involving property; only if a society has the institution of marriage can there be desire- independent reasons involving the institution of marriage. The point, however, which must be emphasized over and over is that the reason does not derive from the institution, rather the institution provides the framework, the structure, within which one creates the reason. But the reason derives from the fact that the agent binds the will through a free and voluntary act.

Rationality

In order that the practice of creating desire-independent reasons can ever be socially effective, it must be effective in virtue of the rationality of the agents involved. It is only because I am a rational agent that I can recognize that my previous behavior has created reasons for my present behavior.

Combining all five elements

Now let us try to put these points together into a general account. To begin with, how can we organize time? The obvious answer is that we do things now which will make things happen in the future in a way they would not have happened if we did not act now. That is why we set our alarm clocks. We know we have a reason to get up at 6 a.m., but we also know that at 6 a.m. we will not be able to act on that reason because we will be asleep. So, by setting the alarm clock now, we will make it possible to act on a reason in the future. But suppose I don't have an alarm clock and I have to try to get some other person to wake me up. What is the difference between setting an alarm clock for 6 a.m. and asking someone to wake me up at 6 a.m., for example? In both cases I do something now to make it the case that I will wake up at 6 a.m. tomorrow. The difference is that in the former case only causes are created, in the latter case new reasons for action are created. How? Well, there are different sorts of cases. If I don't trust the person in question I might say, "If you wake me up at 6 a.m. I will give you five dollars." In that case I have made a promise, a conditional promise to give the other person five dollars; and, if he accepts the offer, he has promised to wake me up on the condition that I pay him five dollars. This is typical of contracts. Each party makes a promise, conditional on receiving a benefit from the other party.

In the more realistic case I simply extract from him a promise to wake me up. I say, "Please wake me up at 6 a.m.," and he says, "OK"; in that context he has made an unconditional promise and created a desire-independent reason.

In a third sort of case, no promise need be made at all. Suppose I do not trust the person at all, but I know that he makes his breakfast every day at 6 a.m. I simply position all the breakfast food so that he can't get at it without waking me up. I take it in my room and lock the door, for example. To get breakfast he has to bang on my door to wake me up. Now this third sort of case also creates a reason to wake me up, but this one is a prudential or desire-dependent reason. He has to reason: "I want breakfast, I can't have breakfast unless I wake him up, so I will wake him up."

All three of these methods might on occasion work equally well in getting someone else to wake me up, but I want to call attention to what a bizarre case the third one is. If the only way we could get cooperation from other people was by getting them in a position so that they independently of us want to do what we want them to do, most forms of human social life would be impossible. *In order that we can organize time on a social basis it is necessary that we create mechanisms to justify reasonable expectations about the future behavior of members of the community, ourselves included.* The reason it is hard to do that is that every intentional action is the expression of a desire, and our problem is to arrange things so that people will have the appropriate desires at the appropriate times. But it is impossible to do that if all we had to go on are desire-dependent reasons—on preexisting desires, inclinations, etc. There would have to be a miraculous preestablished harmony between our desires and other people's desires. We would have to arrange things now so that we and our conspecifics would want to do such and such in the future.

The only way we can organize time is to create a set of mechanisms whereby we and our fellow beasts can create reasons now for doing something in the future, regardless of whether we feel like doing it when the time comes. But how is that possible?

Let us turn to the role of language and other institutional structures. There are many features of institutional facts that require analysis, and I have elsewhere tried to give an analysis of several of them and I won't repeat it here (Searle 1995). However, there is one feature that is essential for the present discussion. In the case of institutional facts, the normal relationship between intentionality and ontology is reversed. In the normal case, what *is* the case is logically prior to what *seems to be* the case. So, we understand that the object seems to be heavy, because we understand what it is for an object to be heavy. But in the case of institutional reality, the ontology derives from the intentionality. In order for something to be money, people have to think that it is money. But if enough of them think it is money and have other appropriate attitudes, and act appropriately, then it is money. If we all think that a certain sort of thing is money and we cooperate in using it, regarding it, treating it as money, then it is money. In this case, "seems" is prior to "is". I cannot exaggerate the importance of this phenomenon. The noises coming out of my mouth, seen as part of physics, are rather trivial acoustic blasts. But they have remarkable features. Namely, we think they are sentences of English and that their utterances are speech acts. If we all think that of them as sentences and speech acts, and if we all cooperate in using, interpreting, regarding, responding to, and generally treating them as sentences and speech acts, then they are what we use, regard, treat, and interpret. (I am being very brief here. I do not wish to suggest that these phenomena are in any way simple.) In such cases we create an institutional reality by treating a brute reality as having a certain status. The entities in question—money, property, government, marriage, universities, and speech acts—all have a level of description where they are brute physical phenomena like mountains and snowdrifts. But by collective intentionality we impose on them statuses, and with those statuses functions that they cannot perform without that imposition.

The next step is to see that in the creation of these institutional phenomena we can also create reasons for action. I have a reason for preserving and maintaining the rather sordid bits of paper in my wallet, because I know that they are more than just bits of paper. They are valuable pieces of United States currency. That is, given the institutional structure, there are whole sets of reasons for actions which could not exist without the institutional structure. So, "it seems to be the case" can create a set of reasons for action, because what seems to be the case (appropriately understood) is the case, where institutional reality is concerned. If I borrow money from somebody, or order a beer in a bar, or get married, or join a club, I use institutional structures to create reasons for action and the reasons exist within institutional structures.

But so far this doesn't answer our crucial question, namely, how can we use such structures to create desire-independent reasons? I have very good reasons for wanting money, but they are all desire-dependent, because they derive from the desires I have for the things I can buy with the money. But what about the obligations I have to pay money? Or my debts to other people? Or my promises to deliver money on such and such occasions? If a group of people create an institution whose sole function is that I should pay them money, I have so far no obligation whatever to pay them money, because though they might have created what they think is a reason, it is not yet a reason for me. So, how can I use institutional reality to create desire-independent reasons for me?

It is at this point that we have to introduce the features of freedom and the first-person point of view. Our question now is, How can I create a reason for myself, a reason which will be binding on me in the future, even though I may not at that time have any desire to do the thing I created a reason for doing. I think the question becomes impossible to answer if you look at the phenomena from the third-person point of view. From a third-person point of view, someone makes a bunch of noises through his mouth. He says, "I promise to wake you up at 6 a.m." How can doing that ever create a reason which will bind his will? The only way to answer this question is to see from the first-person point of view what I think is going on, what I am trying to do, what is my intention, when I make these sounds through my mouth. And once we see the matter from the first-person point of view we can, I believe, see the solution to our puzzle. When I say "I promise to wake you at 6 a.m." I see myself as freely creating a special type of desire-independent reason, an obligation, for me to wake you at 6 a.m. This is the whole point of promising. Indeed, that is what a promise is. It is the intentional creation of certain sort of obligation—and such obligations are by definition independent of the subsequent desires of the agent. But now, all I have said so far is that I made noises with certain intentions and, because I have those intentions, such and such seems to me to be the case. But how do we get from "it seems to be the case" to "it is the case", and to answer that question, we have to go back to what I just said about institutional structures. It is characteristic of these structures that *seems* is prior to *is*. If it seems to me that I am creating a promise, because that was my intention in doing what I did, and it seems to you that you have received a promise, and all of the other conditions which I will not enumerate here but have enumerated in tedious detail elsewhere (Searle 1969), if all the other conditions on the possibility of creating a promise are present, then I have created a promise. I have intentionally created a new entity, such that it is binding on me in the future; that is, it is a desire-independent reason for me because I have freely and intentionally created it as such.

The ability to bind the will now creates a reason for the future act only because it is a manifestation of my freedom now. I said earlier that this shows why the slave doesn't have any reason to obey the slave owner, except desire-dependent reasons, even though both promissor and slave act within institutional structures. The only reasons the slave has are prudential reasons. The slave never exercised any freedom in creating a reason for himself to act. To see how within the institutional structure an agent can create external reasons for acting, it is essential to see that within the institutional structure there is the possibility of the agent freely creating reasons for himself. There cannot be any question that it is a reason for him because he has freely and voluntarily created it as a reason for him. Now, this is not to say, of course, that it is a reason that will override all other reasons. On the contrary, in any real-life situation, there will be a large number of competing reasons for doing any action, or for not doing that action. When the time comes, the agent still may have to weigh his promise against all sorts of other competing reasons for doing or not doing the action.

We have so far considered four features: time, institutional structures, the first-person point of view, and freedom. I now turn to the fifth, rationality. The ability to act rationally is a general set of capacities involving such things as the ability to recognize and operate with consistency, inference, recognition of evidence and a large number of others. The features of rationality that are important for the present discussion involve the capacity to operate in various ways with reasons for action. I want that to sound vague at this point because clarifying it is our next essential task.

Suppose I have freely acted with the intention of creating a desire-independent reason for me, suppose I have met all the conditions (on promising, or ordering a beer, or whatever), so that I really succeeded in creating that reason, then when the time comes, what do I need in order to recognize that there is such a reason? Assuming that I know all the facts, rationality is sufficient for recognizing that the prior creation of a reason is now binding. The important thing is that you don't have to have some extra moral principle about promising or beer drinking in order to understand that the reason you created in the past as a binding reason for the present moment is precisely a binding reason in the present moment. It is sheer logical inconsistency to grant all the facts, about the creation and continuation of the obligation, and then to deny that you have a reason for acting. That was the point of the absurd example with which I began. Forget for a moment about what is "moral" and "immoral": the point is that it is *logically* absurd and therefore *irrational* to say, as our imagined speaker did, "I accept that I made a promise and that promises are undertakings of obligations, and I accept that I have no conflicting obligations or other reasons for not keeping my promise, but what I don't see is how any of that gives me any reason at all to do the thing I promised to do."

4. Summary and Conclusion

In this essay I have been concerned to show how human beings can create desire-independent reasons for action. What facts correspond to the claim that the agent has created such a reason? (1) There must exist a structure suffcent for the creation of such institutional facts. These structures are invariably linguistic but they may involve other institutions as well. Such structures enable us to buy a house, order a beer, enroll in college, etc. (2) Within these structures, if the agent acts with the appropriate intentions, that is sufficient for the creation of desire-independent reasons. Specifically, if the agent acts with the intention that his action should create such a reason, then, if the circumstances are otherwise appropriate, he has created such a reason. The crucial intention is the intention by the agent to bind his own will. The reason derives from the agent's intention that it be a reason, not from the institution alone. The institution provides only the vehicle for the creation of the reason.

Once created, it is a requirement of rationality that the agent should recognize the reason created as binding on his subsequent behavior.

I do not, of course, say that these are the only forms of desire-independent reasons. Perhaps there are others. But I believe these are sufficient to refute the Classical Model on this issue.

Bibliography

ACTON, B. (1993), "A Network Model of Visual Indexing and Attention", M.Sc. thesis, Department of Electrical Engineering, University of Western Ontario.

ADOLPHS, R., and DAMASIO, A. R. (1998), "The Human Amygdala in Social Judgment", *Nature*.

——— TRANEL, D., DAMASIO, H., and DAMASIO, A. R. (1994), "Impaired Recognition of Emotion in Facial Expressions Following Bilateral Damage to the Human Amygdala", *Nature*, 372: 669–72.

——————————— (1995), "Fear and the Human Amygdala", *Journal of Neuroscience*, 15: 5879–91.

——— ——— BECHARA, A., DAMASIO, H., and DAMASIO, A. R. (1996), "Neuropsychological Approaches to Reasoning and Decision Making", in A. R. Damasio, et al. (eds.), *The Neurobiology of Decision-Making* (Berlin: Springer-Verlag), 157–80.

AGGLETON, J. P., and PASSINGHAM, R. E. (1981), "Syndrome Produced by Lesions of the Amygdala in Monkeys *(Macaca mulata)*", *Journal of Comparative and Physiological Psychology*, 95: 961–77.

ALBERT, M. S., BUTTER, N., and LEVIN, J. A. (1979), "Temporal Gradients in the Retrograde Amnesia of Patients with Alcoholic Korsakoff's Disease", *Archives of Neurology*, 36: 211–16.

ALOIMONOS, Y. (1993), *Active Perception* (Hillsdale, NJ: Lawrence Erlbaum Associates).

ANDERSEN, R. (1995), "Encoding of Intention and Spatial Location in the Posterior Parietal Cortex", *Cerebral Cortex*, 5: 457–69.

——— ESSICK, G., and SIEGEL, R. (1985), "Encoding of Spatial Location by Posterior Parietal Neurons", *Science*, 230: 456–8.

——— SNYDER, L., BRADLEY, D., and XING, J. (1997), "Multimodal Representation of Space in the Posterior Parietal Cortex and its Use in Planning Movements, in W. Cowan, E. Shooter, C. Stevens, and R. Thompson (eds.), *Annual Review of Neuroscience*, 20: 303–30.

——— BRACEWELL, R., BARASH, S., GNADT, J., and FOGASSI, L. (1990), "Eye Position Effects on Visual, Memory, and Saccade-Related Activity in Areas LIP and 7a of Macaque", *Journal of Neuroscience*, 10: 1176–96.

ANTELL, S., and KEATING, D. (1983), "Perception of Numerical Invariance in Neonates", *Child Development*, 54: 695–701.

ANTINUCCI, F. (1989), *Cognitive Structure and Development in Nonhuman Primates* (Hillsdale, NJ: Lawrence Erlbaum Associates).

BAILLARGEON, R. (1987), "Young Infant's Reasoning about the Physical and Spatial Properties of a Physical Object", *Cognitive Development*, 2: 179–200.

——— (1991), "Reasoning about the Height and Location of a Hidden Object in 4.5 and 6.5 Month Old Infants", *Cognition*, 38: 13.

——— (1994), "A Model of Physical Reasoning in Infancy", in C. Rovee-Collier and L. Lipsitt (eds.), *Advances in Infancy Research*, ix (Norwood, NJ: Ablex), 114–39.

——— and DEVOS, J. (1991), "Object Permanence in Young Infants: Further Evidence", *Child Development*, 62: 1227–46.

BAILLARGEON, R. SPELKE, E., and WASSERMAN, S. (1985), "Object Permanence in 5-month-old Infants", *Cognition*, 20: 191–208.

BALLARD, D. (1991), "Animate Vision", *Artificial Intelligence*, 48: 57–86.

BALLARD, D. H., HAYHOE, M. M., POOK, P. K., and RAO, R. P. N. (1997), "Deictic Codes for the Embodiment of Cognition", *Behavioral and Brain Sciences*, 20: 723–67.

BARD, P. (1928), "A Diencephalic Mechanism for the Expression of Rage with Special Reference to the Sympathetic Nervous System", *American Journal of Physiology*, 84: 490–515.

BARKOW, J., COSMIDES, L., and TOOBY, J. (eds.) (1992), *The Adapted Mind: Evolutionary Psychology and the Generation of Culture* (Oxford: Oxford University Press).

BAYLIS, G. C., and DRIVER, J. (1993), "Visual Attention and Objects: Evidence for Hierarchical Coding of Location", *Journal of Experimental Psychology: Human Perception and Performance*, 19: 451–70.

BECHARA, A., DAMASIO, A. R., DAMASIO, H., and ANDERSON, S. W. (1994), "Insensitivity to Future Consequences Following Damage to Human Prefrontal Cortex", *Cognition*, 50: 7–15.

BEDAU, M. A. (1996), "The Nature of Life", in M. A. Boden (ed.), *The Philosophy of Artificial Life* (Oxford: Oxford University Press), 332–57.

BICKERTON, D. (1990), *Species and Language* (Chicago: University of Chicago Press).

BISIACH, E. and LUZZATTI, C. (1978), "Unilateral Neglect of Representational Space", *Cortex*, 14: 129–33.

BLOCK, N. (1995), "On a Confusion about a Function of Consciousness", *Behavioral and Brain Sciences*, 18/2: 227–47.

—— (1996), Review of Francis Crick, *The Astonishing Hypothesis*, *Contemporary Psychology*, May.

BODEN, M. A. (1990), "Escaping from the Chinese Room", in M. A. Boden (ed.), *The Philosophy of Artificial Intelligence* (Oxford: Oxford University Press), 89–104.

—— (1996), "Autonomy and Artificiality", in M. A. Boden (ed.), *The Philosophy of Artificial Life* (Oxford: Oxford University Press), 95–108.

—— (1998), "Consciousness and Human Identity: An Interdisciplinary Perspective", in J. Cornwell (ed.), *Consciousness and Human Identity* (Oxford: Oxford University Press), 1–20.

—— (1999), "Is Metabolism Necessary?", *British Journal for the Philosophy of Science*.

BOGEN, J. E., and BOGEN, G. M. (1969), "The Other Side of the Brain, III: The Corpus Callosum and Creativity", *Bulletin of the Los Angeles Neurology Society*, 34: 191–220.

BOROD, J. C. (1992), "Interhemispheric and Intrahemispheric Control of Emotion: A Focus on Unilateral Brain Damage", *Journal of Consulting and Clinical Psychology*, 60: 339–48.

BOYSEN, S. T. (1993), "Counting in Chimpanzees: Nonhuman Principles and Emergent Properties of Number", in S. T. Boysen and E. J. Capaldi, (eds.), *The Development of Numerical Competence: Animal and Human Models*, (Hillsdale, NJ: Lawrence Erlbaum Associates), 39–59.

—— and CAPALDI, E. J. (1993), *The Development of Numerical Competence: Animal and Human Models*, (Hillsdale, NJ: Lawrence Erlbaum Associates).

BROCA, P. (1861), "Remarques sur le siège de la faculté de langage articulé, suivies d'une observation d'aphémie (perte de la parole)", *Bulletin de la Societé d'Anatomie*, 36: 330–57.

BROOKS, R. A. (1991), "Intelligence without Representation", *Artificial Intelligence*, 47: 139–59.

BROWN, C. M. (1984), "Computer Vision and Natural Constraints", *Science*, 224: 1299–305.

BURGE, T. (1986), "Cartesian Error and the Objectivity of Perception", in Philip Pettit and John McDowell (eds.), *Subject, Thought, and Context* (Oxford: Oxford University Press), 117–36.

—— (1988), "Individualism and Self-Knowledge", *Journal of Philosophy*, 85: 649–63.

BURKELL, J., and PYLYSHYN, Z. W. (1997), "Searching through Subsets: A Test of the Visual Indexing Hypothesis", *Spatial Vision*, 11: 225–58.

BURKS, A. W. (ed.), (1970), *Essays on Cellular Automata* (Champaign-Urbana: University of Illinois Press). (Posthumously edited and completed essays of J. von Neumann.)

BUTLER, A., and HODOS, W. (1996), *Comparative Vertebrate Neuroanatomy: Evolution and Adaptation* (New York: Wiley-Liss).

CAPALDI, E. J. (1993), "Animal Number Abilities: Implications for a Hierarchical Approach to Instrumental Learning", in S. T. Boysen and E. J. Capaldi, (eds.), *The Development of Numerical Competence: Animal and Human Models*, (Hillsdale, NJ: Lawrence Erlbaum Associates), 191–209.

CAREY, S. (1995), "Continuity and Discontinuity in Cognitive Development", in E. Smith and D. Osherson (eds.), *Thinking*, vol. iii of *An Invitation to Cognitive Science*, 2nd edn. (Cambridge, Mass: MIT Press), 101–30.

CHAN, W. P., PRETE, F., and DICKINSON, M. H. (1998), "Visual Input to the Efferent Control System of a Fly's 'Gyroscope'", *Science*, 280: 289–92.

CHASTAIN, G. (1995), "Location Coding with Letters versus Unfamiliar, Familiar and Labeled letter-like forms", *Canadian Journal of Experimental Psychology*, 49: 95–112.

CHENEY, D., and SEYFARTH, R. (1990), *How Monkeys see the World: Inside the Mind of Another Species* (Chicago: University of Chicago Press).

CHOMSKY, N. (1965), *Aspects of the Theory of Syntax* (Cambridge, Mass.: MIT Press).

—— (1986), *Knowledge of Language: Its Nature, Origin, and Use* (New York: Praeger).

CHURCH, R. M., and BROADBENT, H. A. (1990), "Alternative Representations of Time, Number, and Rate", *Cognition*, 37: 55–81.

CHURCHLAND, P. M. (1989a), *A Neurocomputational Perspective: The Nature of Mind and the Structure of Science* (Cambridge, Mass.: MIT Press).

—— (1989b), "On the Nature of Theories: A Neurocomputational Perspective", in W. Savage (ed.), *Scientific Theories*, Minnesota Studies in the Philosophy of Science, 14 (Minneapolis: University of Minnesota Press), 59–101. (Ch. 9 of Churchland 1989a.)

—— (1989c), "On the Nature of Explanation: A PDP Approach", ch. 10 of Churchland 1989a. Reprinted in J. Misiek (ed.), *Rationality*, Boston Studies in the Philosophy of Science, 175 (Dordrecht: Kluwer, 1995).

—— (1989d), "Learning and Conceptual Change", ch. 11 of Churchland 1989a.

—— (1995), *The Engine of Reason, the Seat of the Soul: A Philosophical Journey into the Brain* (Cambridge, Mass.: MIT Press).

CHURCHLAND, P. S., and SEJNOWSKI, T. J. (1992), *The Computational Brain* (Cambridge, Mass.: MIT Press).

CLARK, A. J. (1993), *Associative Engines: Connectionism, Concepts and Representational Change* (Cambridge, Mass.: MIT Press).

CLARK, A. J. (1997), *Being There: Putting Brain, Body, and World Together Again* (Cambridge, Mass.: MIT Press).

CLIFF, D., and MILLER, G. F. (1995), "Co-Evolution of Pursuit and Evasion II: Simulation Methods and Results", *Adaptive Behavior.* (An abridged version is available in P. Maes, M. J. Mataric, J.-A. Meyer, J. Pollack, and S. W. Wilson (eds.), *From Animals to Animats 4: Proceedings of the Fourth International Conference on Simulation of Adaptive Behavior* (SAB96) (Cambridge, Mass.: MIT Press, 1996), 506–15.

—— HARVEY, I., and HUSBANDS, P. (1993), "Explorations in Evolutionary Robotics", *Adaptive Behavior,* 2: 71–108.

COLBY, C. and DUHAMEL, J. (1993), "Ventral Intraparietal Area of the Macaque: Anatomic Location and Visual Response Properties", *Journal of Neurophysiology,* 69: 902–14.

COOPER, L. (1982), "Demonstration of a Mental Analog of an External Rotation", in R. Shepard and L. Cooper, *Mental Images and their Transformations* (Cambridge, Mass.: MIT Press).

—— and SHEPARD, R. (1982), "Chronometric Studies of the Rotation of Mental Images", in R. Shepard and L. Cooper, *Mental Images and their Transformations* (Cambridge, Mass.: MIT Press).

COSMIDES, L., and TOOBY, J. (1994*a*), "Beyond Intuition and Instinct Blindness: Toward an Evolutionarily Rigorous Cognitive Science", *Cognition,* 50: 41–77.

—— —— (1994*b*), "Origins of Domain Specificity: The Evolution of Functional Organization", in L. A. Hirschfeld and S. A. Gelman (eds.), *Mapping the Mind: Domain Specificity in Cognition and Culture* (New York: Cambridge University Press), 85–116.

COTTRELL, G. (1991), "Extracting Features from Faces Using Compression Networks: Face, Identity, Emotions and Gender Recognition Using Holons", in D. Touretzky, J. Elman, T. Sejnowski, and G. Hinton (eds.), *Connectionist Models: Proceedings of the 1990 Summer School.* San Mateo, (Calif.: Morgan Kaufmann).

CRICK, F. (1994), *The Astonishing Hypothesis* (New York: Scribner's).

—— (1996), "Visual Perception: Rivalry and Consciousness", *Nature,* 380: cf Damasio *et al.* 1996, 485–6.

—— and KOCH, C. (1995*a*), "Are We Aware of Neural Activity in Primary Visual Cortex?", *Nature,* 375: 121–3.

—— —— (1995*b*), "Why Neuroscience May be Able to Explain Consciousness", sidebar in *Scientific American,* 12/95: 92.

—— —— (1995*c*), Untitled response to Pollen, *Nature,* 377: 294–5.

DAMASIO, A. R. (1994), *Descartes's Error: Emotion, Reason and the Human Brain* (New York: Grosset/Putnam).

—— (1995), "Toward a Neurobiology of Emotion and Feeling: Operational Concepts and Hypotheses", *Neuroscientist,* 1: 19–25.

—— (1996), "The Somatic Marker Hypothesis and the Possible Functions of the Prefrontal Cortex", *Transactions of the Royal Society,* 351: 1413–20.

—— (1999), *The Feeling of What Happens: Body and Emotion in the Making of Consciousness* (New York: Harcourt Brace).

—— and DAMASIO, H. (1989), "The Brain Binds Entities and Events by Multiregional Activation from Convergence Zones", *Neural Computation,* 1: 123–32.

—— —— (1992), "Brain and Language", *Scientific American,* 267: 33–95.

———— (1994), "Cortical Systems for Retrieval of Concrete Knowledge: The Convergence Zone Framework", in C. Koch (ed.), *Large-Scale Neuronal Theories of the Brain* (Cambridge, Mass.: MIT Press), 61–74.

—— DAMASIO, A. R., GRABOWSKI, T. J., BECHARA, A., DAMASIO, H., PONTO, L. L. B., and HICHWA, R. D. (1998), "Neural Correlates of the Experience of Emotions", *Society for Neuroscience*, 24: 258.

—— and TRANEL, D. (1993), "Nouns and Verbs are Retrieved with Differently Distributed Neural Systems", *Proceedings of the National Academy of Sciences*, 90: 4957–60.

—— and VAN HOESEN, G. W. (1983), "Emotional Disturbances Associated with Focal Lesions of the Limbic Frontal Lobe", in K. M. Heilman and P. Satz (eds.), *The Neuropsychology of Human Emotion: Recent Advances* (New York: Guilford Press), 85–110.

—— TRANEL, D., and DAMASIO, H. (1991), "Somatic Markers and the Guidance of Behavior: Theory and Preliminary Testing", in H. S. Levin, H. M. Eisenberg, and A. L. Benton (eds.), *Frontal Lobe Function and Dysfunction* (New York: Oxford University Press), 217–29.

—— DAMASIO, H., TRANEL, D., and BRANDT, J. P. (1990), "Neural Regionalization of Knowledge Access: Preliminary Evidence", *Symposia on Quantitative Biology*, 55: 1039–47.

DAMASIO, H. (1995), *Human Brain Anatomy in Computerized Images* (New York: Oxford University Press).

—— and DAMASIO, A. R. (1989), *Lesion Analysis in Neuropsychology* (New York: Oxford University Press).

—— and FRANK, R. (1992), "Three-Dimensional *in vivo* Mapping of Brain Lesions in Humans", *Archives of Neurology*, 49: 137–43.

—— GRABOWSKI, T. J., TRANEL, D., HICHWA, R., and DAMASIO, A. R. (1996), "A Neural Basis for Lexical Retrieval", *Nature*, 380: 499–505.

———— FRANK, R. J., KNOSP, B., HICHWA, R. D., WATKINS, G. L., and PONTO, L. L. (1993), "PET-Brainvox, a Technique for Neuroanatomical Analysis of Positron Emission Tomography Images", in K. Uemura, N. A. Lassen, T. Jones, and I. Kanno (eds.), *Quantification of Brain Function: Tracer Kinetics and Image Analysis in Brain PET* (Amsterdam: Elsevier Science Publishers), 465–73.

DARWIN, C. (1859), *On the Origin of Species* (London: John Murray).

—— (1871), *The Descent of Man and Selection in Relation to Sex* (London: John Murray).

DAVIDSON, D. (1967), "Truth and Meaning", in D. Davidson, *Inquiries into Truth and Interpretation* (Oxford: Clarendon Press, 1984), 17–36.

—— (1986), "A Nice Derangement of Epitaphs", in E. LePore (ed.), *Truth and Interpretation: Perspectives on the Philosophy of Donald Davidson.* (Oxford: Basil Blackwell), 433–46.

—— (1990), "The Structure and Content of Truth", *Journal of Philosophy*, 87: 279–328.

DAVIDSON, R. (1992), "Prolegomenon to Emotion: Gleanings from Neuropsychology", *Cognition and Emotion*, 6: 245–68.

DAVIS, H., and PERUSSE, R. (1988), "Numerical Competence in Animals: Definitional Issues, Current Evidence, and New Research Agenda", *Behavioural and Brain Sciences*, 11: 561–615.

DAVIS, M. (1992), "The Role of the Amygdala in Conditioned Fear", in J. P. Aggleton (ed.), *The Amygdala: Neurobiological Aspects of Emotion, Memory, and Mental Dysfunction* (New York: Wiley-Liss), 255–302.

DAWKINS, R. (1986), *The Blind Watchmaker* (New York: Norton).

DAWSON, M., and PYLYSHYN, Z. W. (1988), "Natural Constraints in Apparent Motion", in Z. W. Pylyshyn (ed.), *Computational Processes in Human Vision: An Interdisciplinary Perspective* (Norwood, NJ: Ablex), 99–120.

DEACON, T. W. (1997), *The Symbolic Species* (New York: Norton).

DENNETT, D. C. (1987), *The Intentional Stance* (Cambridge, Mass.: MIT Press).

——(1994), "The Practical Requirements for Making a Conscious Robot", *Transactions of the Royal Society*, series A, 349: 133–46. (Special issue, *Artificial Intelligence and the Mind: New Breakthroughs or Dead-Ends?*, ed. M. A. Boden, A. Bundy, and R. M. Needham.)

——(1996), *Kinds of Minds* (New York: Basic Books).

DE SOUSA, R. (1991), *The Rationality of Emotion* (Cambridge, Mass.: MIT Press).

DIAMOND, A. (1988), "Differences between Adult and Infant Cognition: Is the Crucial Variable Presence or Absence of Language?", in L. Weiskrantz (ed.), *Thought without Language* (Oxford: Clarendon Press).

——(1991), "Neuropsychological Insights into the Meaning of the Object Concept Development", in S. Carey and R. Gelman, (eds.), *The Epigenesis of Mind* (Hillsdale, NJ: Lawrence Erlbaum Associates), 112–34.

DORÉ, F. Y., and DUMAS, C. (1987), "Psychology of Animal Cognition: Piagetian Studies", *Psychology Bulletin*, 102: 219–33.

DRETSKE, F. (1988), *Explaining Behavior* (Cambridge, Mass.: MIT Press).

——(1994), "The Explanatory Role of Information", *Transactions of the Royal Society*, series A, 349: 59–70. (Special issue, *Artificial Intelligence and the Mind: New Breakthroughs or Dead-Ends?*, ed. M. A. Boden, A. Bundy, and R. M. Needham.)

DRIVER, J., and HALLIGAN, P. (1991), "Can Visual Neglect Operate in Object-Centered Coordinates? An Affirmative Single Case Study", *Cognitive Neuropsychology*, 9: 475–94.

DUMMETT, M. (1993), *Origins of Analytical Philosophy* (Cambridge, Mass.: Harvard University Press).

DUNCAN, J. (1984), "Selective Attention and the Organization of Visual Information", *Journal of Experimental Psychology: General*, 113: 501–17.

EILENBERG, S. (1974), *Automata, Languages, and Machines* (New York: Academic Press).

EKMAN, P. (1993), "Facial Expressions and Emotion", *American Psychologist*, 48: 348–92.

ELMAN, J. L. (1992), "Grammatical Structure and Distributed Representations", in S. Davis (ed.), *Connectionism: Theory and Practice.* Vancouver Studies in Cognitive Science, 3 (Oxford: Oxford University Press).

——BATES, E., JOHNSON, M., KARMILOFF-SMITH, A., PARISI, D., and PLUNKETT, K. (1996), *Rethinking Innateness: A Connectionist Perspective on Development* (Cambridge, Mass.: MIT Press).

ELMAN, J., and ZIPSER, D. (1988), "Learning the Hidden Structure of Speech", *Journal of the Acoustical Society of America*, 83: 1615–26.

ERIKSEN, C. W., and ST JAMES, J. D. (1986), "Visual Attention Within and Around the Field of Focal Attention: A Zoom Lens Model", *Perception and Psychophysics*, 40: 225–40.

EVANS, C. S., and MARLER, P. (1995), "Language and Communication: Parallels and Contrasts", in H. L. Roitblat and J.-A. Meyer (eds.), *Comparative Approaches to Cognitive Science* (Cambridge, Mass.: MIT Press), 341–82.

FARAH, M. J. (1990), *Visual Agnosia: Disorders of Object Recognition and What they Tell us about Normal Vision* (Cambridge, Mass.: MIT Press).

FELDMAN, J. A., and BALLARD, D. H. (1982), "Connectionist Models and their Properties", *Cognitive Science*, 6: 205–54.

FIELD, H. (1994), "Deflationist Views of Meaning and Content", *Mind*, 103: 249–85.

FINLAY, B. L., and DARLINGTON, R. B. (1995), "Linked Regularities in the Development and Evolution of Mammalian Brains", *Science*, 258: 1578–84.

—— HERSMAN, M., and DARLINGTON, R. B. (1998), "Patterns of Vertebrate Neurogenesis and the Paths of Vertebrate Evolution", *Brain, Behavior and Evolution*, 52: 232–42.

FIVUSH, R., and HAMOND, N. (1990), "Autobiographical Memory across the Preschool Years: Toward Reconceptualizing Childhood Amnesia", in R. Fivush and J. Hudson (eds.), *Knowing and Remembering in Young Children* (New York: Cambridge University Press).

FLANAGAN, O. (1991), *Varieties of Moral Personality: Ethics and Psychological Realism* (Cambridge, Mass.: Harvard University Press).

—— (1996), "The Moral Network", in B. McCauley (ed.), *The Churchlands and their Critics* (Cambridge, Mass.: Blackwell Publishers), 192–215.

FODOR, J. A. (1975), *The Language of Thought* (New York: Crowell).

—— (1983), *The Modularity of Mind* (Cambridge, Mass.: MIT Press/Bradford Books).

—— (1990), *A Theory of Content and Other Essays* (Cambridge, Mass.: MIT Press).

—— (1995), "The Folly of Simulation", in P. Baumgartner and S. Payr (eds.), *Speaking Minds* (Princeton: Princeton University Press).

—— (1998), *Concepts: Where Cognitive Science Went Wrong.* (Oxford: Oxford University Press).

FOGASSI, L., GALLESE, V., DI PELLEGRINO, G., FADIGA, L., GENTILUCCI, M., LUPPINO, G., MATELLI, M., PEDOTTI, A., and RIZZOLATTI, G. (1992), "Space Coding by Premotor Cortex", *Experimental Brain Research*, 89: 686–90.

FRANCIS, W. N., and KUCERA, H. (1982), *Frequency Analysis of English Usage: Lexicon and Grammar* (Boston: Houghton Mifflin).

FRANK, R. J., DAMASIO, H., and GRABOWSKI, T. J. (1997), "Brainvox: An Interactive, Multimodal, Visualization and Analysis System for Neuroanatomical Imaging", *NeuroImage*, 3: 13–30.

FRISTON, K. J., HOLMES, P. P., WORSLEY, K. J., POLINE, M.-B., FRITH, C. D., and FRACKOWIAK, R. S. J. (1995), "Statistical Parametric Maps in Functional Imaging: A General Linear Approach", *Human Brain Mapping*, 2: 189–210.

FUSON, K. C. (1988), *Children's Counting and Concepts of Number* (New York: Springer-Verlag).

GAINOTTI, G. (1972), "Emotional Behavior and Hemispheric Side of the Lesion", *Cortex*, 8: 41–55.

GALLISTEL, C. R. (1990), *The Organization of Learning* (Cambridge, Mass: MIT Press).

GARDNER, H., BROWNELL, H. H., WAPNER, W., and MICHELOW, D. (1983), "Missing the Point: The Role of the Right Hemisphere in the Processing of Complex

Linguistic Materials", in E. Pericman (ed.), *Cognitive Processes and the Right Hemisphere* (New York: Academic Press).

GELMAN, R. (1990), "First Principles Organize Attention to and Learning about Relevant Data: Number and the Animate–Inanimate Distinction as Examples", *Cognitive Science*, 14: 79–106.

—— and GALLISTEL, C. R. (1978), *The Child's Understanding of Number* (Cambridge, Mass.: Harvard University Press).

GEORGE, A. (1989), "How not to Become Confused about Linguistics", in A. George (ed.), *Reflections on Chomsky* (Oxford: Basil Blackwell), 90–110.

GESCHWIND, N. (1965), "Disconnexion Syndromes in Animals and Man", *Brain*, 88: 270–94.

GIROSI, F., JONES, M., and POGGIO, T. (1995), "Regularization Theory and Neural Networks Architectures", *Neural Computation*, 7: 219–69.

GOODMAN, N. (1976), *Languages of Art* (Indianapolis: Hackett).

GOODWIN, B. (1994), *How the Leopard Changed its Spots: The Evolution of Complexity* (London: Weidenfeld and Nicolson).

GORMAN, R. P., and SEJNOWSKI, T. J. (1988), "Analysis of Hidden Units in a Layered Network Trained to Classify Sonar Targets", *Neural Networks*, 1: 75–89.

GOUZOULES, S., GOUZOULES, H., and MARLER, P. (1984), "Rhesus Monkey (*Macaca mulatta*) Screams: Representational Signalling in the Recruitment of Agonistic Aid", *Animal Behaviour*, 32: 182–93.

GRABOWSKI, T. J., DAMASIO, H., and DAMASIO, A. R. (1998), "Premotor and Prefrontal Correlates of Category-Related Lexical Retrieval", *NeuroImage*, 7: 232–43.

—— —— FRANK, R., HICHWA, R. D., BOLES-PONTO, L. L., and WATKINS, G. L. (1995), "A New Technique for PET Slice Orientation and MRI-PET Coregistration", *Human Brain Mapping*, 2: 123–33.

—— —— —— BROWN, C. K., BOLES-PONTO, L. L., WATKINS, G. L., and HICHWA, R. D. (1995), "Neuroanatomical Analysis of Functional Brain Images: Validation with Retinotopic Mapping", *Human Brain Mapping*, 2: 134–48.

—— —— —— DAMASIO, H., BOLES-PONTO, L. L., WATKINS, G. L., and HICHWA, R. D. (1996), "Reliability of PET Activation across Statistical Methods, Subject Groups, and Sample Sizes", *Human Brain Mapping*, 4: 23–46.

GRAZIANO, M., and GROSS, C. (1993), "A Bimodal Map of Space: Somatosensory Receptive Fields in the Macaque Putamen with Corresponding Visual Receptive Fields", *Experimental Brain Research*, 97: 96–109.

GRICE, P. (1989), *Studies in the Way of Words* (Cambridge, Mass.: Harvard University Press).

GROH, J., and SPARKS, D. (1996), "Saccades to Somatosensory Targets, 3: Eye-Position Dependent Activity in the Primate Superior Colliculus", *Journal of Neurophysiology*, 75: 439–53.

GRUBER, H. E., GIRGUS, J. S., and BANUAZIZI, A. (1971), "The Development of Object Permanence in the Cat", *Developmental Psychology*, 4: 9–15.

GUHRAUER, G. (1846), *Gottfried Wilhelm Freiherr von Leibniz: Eine Biographie* (Breslau).

HALGREN, E. (1992), "Emotional Neurophysiology of the Amygdala within the Context of Human Cognition", in J. P. Aggleton (ed.), *The Amygdala: Neurobiological Aspects of Motion, Memory, and Mental Dysfunction* (New York: Wiley-Liss).

HARTMANN, J. A. *et al.* (1991), "Denial of Visual Perception", *Brain and Cognition*, 16: 29–40.

HARVEY, I. (1992), "Species Adaptation Genetic Algorithms", in F. J. Varela and P. Bourgine (eds.), *Toward a Practice of Autonomous Systems: Proceedings of the First European Conference on Artificial Life* (Cambridge, Mass.: MIT Press), 346–54.

HARVEY, P. H., and PAGEL, M. D. (1991), *The Comparative Method in Evolutionary Biology* (Oxford: Oxford University Press).

HAUSER, M. D. (1996), *The Evolution of Communication* (Cambridge, Mass.: MIT Press/Bradford Books).

—— and CAREY, S. (1998), "Building a Cognitive Creature from a Set of Primitives: Evolutionary and Developmental Insights", in C. Allen and D. Cummings (eds.), *The Evolution of Mind* (Oxford: Oxford University Press), 51–106.

—— MACNEILAGE, P., and WARE, M. (1996), "Numerical Representations in Primates", *Proceedings of the National Academy of Sciences*.

HEILMAN, K., WATSON, R. T., and BOWERS, D. (1983), "Affective Disorders Associated with Hemispheric Disease", in K. Heilman and P. Satz (eds.), *Neuropsychology of Human Emotion* (New York: Guilford Press), 45–64.

HEINRICH, B. (1989), *Ravens in Winter* (New York: Summit Books).

—— (1993), "A Birdbrain Nevermore", *Natural History*, 102: 51–7.

HERRNSTEIN, R. J. (1991), "Levels of Categorization", in G. M. Edelman, W. E. Gall, and W. M. Cowan, (eds.), *Signal and Sense* (New Jersey: Wiley-Liss), 385–413.

HERSCOVITCH, P., MARKHAM, J., and RAICHLE, M. E. (1983), "Brain Blood Flow Measured with Intravenous $H_2^{15}O$, I: Theory and Error Analysis", *Journal of Nuclear Medicine*, 24: 782–89.

HESS, W. R. (1954), *Diencephalon: Autonomic and Extrapyramidal Functions* (New York: Grune and Stratton).

HICHWA, R. D., PONTO, L. L., and WATKINS, G. L. (1995), "Clinical Blood Flow Measurement with [^{15}O] Water and Positron Emission Tomography (PET)", in A. M. Emram (ed.), *Chemists' Views of Imaging Centers* (New York: Plenum Press).

HIGGINBOTHAM, J. T. (1997), "GB Theory: An Introduction", in Johan van Benthem and Alice Meulen (eds), *Handbook of Logic and Language* (Amsterdam: Elsevier), 311–60.

HIRSCH, E. (1982), *The Concept of Identity* (Oxford: Oxford University Press).

HONIG, W. (1993), "Numerosity as a Dimension of Stimulus Control", in S. T. Boysen and E. J. Capaldi (eds.), *The Development of Numerical Competence: Animal and Human Models* (Hillsdale, NJ: Lawrence Erlbaum Associates), 61–86.

HORWICH, P. (1990), *Truth* (Oxford: Basil Blackwell).

—— (1997), "The Composition of Meanings", *Philosophical Review*, 106: 503–32.

INTRILIGATOR, J., and CAVANAGH, P. (1992), "Object-Specific Spatial Attention Facilitation that does not Travel to Adjacent Spatial Locations", *Investigative Opthalmology and Visual Science*, 33: 2849 (abstract).

JAMES W. (1950; orig. pub. 1890), *The Principles of Psychology*, vols. i and ii (New York: Dover).

JEANNEROD, M. (1988), *The Neural and Behavioral Organization of Goal-Directed Movements* (Oxford: Oxford University Press).

JEFFREY, R. (1965), *The Logic of Decision* (Chicago: University of Chicago Press).

JOHNSON, M. (1993), *Moral Imagination* (Chicago: University of Chicago Press).

JOHNSON-LAIRD, P. N. (1978), "The Meaning of Modality", *Cognitive Science*, 2: 17–26.

JOHNSON-LAIRD, P. N. and OATLEY, K. (1992), "Basic Emotions, Rationality, and Folk Theory", *Cognition and Emotion*, 6: 201–23.

KAGAN, J. (1989), *Unstable Ideas: Temperament, Cognition, and Self* (Cambridge, Mass.: Harvard University Press).

KAHNEMAN, D., and TREISMAN, A. (1984), "Changing Views of Attention and Automaticity", in R. Parasuraman and D. R. Davies (eds.), *Varieties of Attention* (New York: Academic Press), 29–61.

KAMP, H., and REYLE, U. (1993), *From Discourse to Logic* (Dordrecht, Holland: Kluwer).

KAUFFMAN, S. A. (1993), *The Origins of Order: Self-Organization and Selection in Evolution* (Oxford: Oxford University Press).

KOCH, C., and ULLMAN, S. (1984), "Selecting One among the Many: A Simple Network Implementing Shifts in Selective Visual Attention", MIT Technology Artificial Intelligence Laboratory, Cambridge, Mass.

KOECHLIN, E., DEHAENE, S., and MEHLER, J. (1996), Numerical Transformations in Five-Month-Old Human Infants", *Mathematical Cognition*, 3: 89–104.

KOSSLYN, S. (1980), *Image and Mind* (Cambridge, Mass.: Harvard University Press).

——(1994), *Image and Brain: The Resolution of the Imagery Debate* (Cambridge, Mass.: MIT Press).

KRIPKE, S. (1980), *Naming and Necessity* (Oxford: Basil Blackwell).

——(1982), *Wittgenstein on Rules and Private Language* (Oxford: Oxford University Press).

KUHL, P. K. (1991), "Perception, Cognition and the Ontogenetic and Phylogenetic Emergence of Human Speech", in S. Brauth, W. Hall, and R. Dooling (eds.), *Plasticity of Development* (Cambridge, Mass.: MIT Press).

LAKOFF, G. (1987), *Women, Fire, and Dangerous Things: What Categories Reveal about the Mind* (Chicago and London: University of Chicago Press).

LANGTON, C. G. (1991), "Life at the Edge of Chaos", in C. G. Langton, C. Taylor, J. D. Farmer, and S. Rasmussen (eds.), *Artificial Life II* (Redwood City, Calif.: Addison-Wesley), 41–91.

LAZARUS, R. S. (1991), *Emotion and Adaptation* (New York: Oxford University Press).

LEDOUX, J. E. (1992), "Emotion and the Amygdala", in J. P. Aggleton (ed.), *The Amygdala: Neurobiological Aspects of Emotion, Memory, and Mental Dysfunction* (New York: Wiley-Liss), 339–51.

——(1993), "Emotional Memory Systems in the Brain", *Behavioural Brain Research*, 58: 69–79.

LEHKY, S., and SEJNOWSKI, T. J. (1988), "Network Model of Shape-from-Shading: Neuronal Function Arises from Both Receptive and Projective Fields", *Nature*, 333: 452–4.

————(1990), "Neural Network Model of Visual Cortex for Determing Surface Curvature from Images of Shaded Surfaces", *Transactions of the Royal Society of London*, series B, 240: 251–78.

LESLIE, A. M. (1982), "The Perception of Causality in Infants", *Perception*, 11: 173–86.

——(1984), "Spatiotemporal Continuity and the Perception of Causality in Infants", *Perception*, 13: 287–305.

——(1994), "ToMM, ToBy, and Agency: Core Architecture and Domain Specificity", in L. A. Hirschfeld and S. A. Gelman (eds.), *Mapping the Mind: Domain Specificity in Cognition and Culture* (New York: Cambridge University Press), 119–48.

——and KEEBLE, S. (1987), "Do six-month old Infants Perceive Causality?", *Cognition*, 25: 265–88.

——XU, F., TREMOLET, P. D., and SCHOLL, B. J. (1998), "Indexing the Object Concept: Developing 'what' and 'where' systems", *Trends in Cognitive Sciences*, 2/1: 10–18.

LEWIS, D. (1986), *On the Plurality of Worlds* (Oxford: Basil Blackwell).

LIEBERMAN, P. (1984), *The Biology and Evolution of Language* (Cambridge, Mass.: Harvard University Press).

LINDSAY, D., JOHNSON, M., and KWON, P. (1991), "Developmental Changes in Memory Source Monitoring", *Journal of Experimental Child Psychology*, 52: 297–318.

LOCKE, J. (1993), *The Path to Spoken Language* (Cambridge, Mass.: Harvard University Press).

LOCKERY, S. R., FANG, Y., and SEJNOWSKI, T. J. (1991), "A Dynamical Neural Network Model of Sensorimotor Transformation in the Leech", *Neural Computation*, 2: 274–82.

LOGAN, C. G., TRANEL, D., FRANK, R., and DAMASIO, A. R. (1996), "Explaining Category-Related Effects in the Retrieval of Conceptual and Lexical Knowledge: Operationalization and Analysis of Factors", *Society for Neuroscience*, 22: 1110.

McCARTHY, J., and HAYES, P. J. (1969), "Some Philosophical Problems from the Standpoint of Artificial Intelligence", in M. Meltzer and D. Michie (eds.), *Machine Intelligence*, iv (Edinburgh: Edinburgh University Press).

McCORMACK, T., and HOERL, C. (1999), "Memory and Temporal Perspective: The Role of Temporal Frameworks in Memory Development", *Developmental Review*, 19: 154–82.

MACINTYRE, A. (1981), *After Virtue* (Notre Dame, Ind.: University of Notre Dame Press).

McKEEVER, P. (1991), "Nontarget Numerosity and Identity Maintenance with FINSTs: A Two Component Account of Multiple Object Tracking", MA diss., University of Western Ontario.

MACLEAN, P. D. (1970), "The Triune Brain, Emotion, and Scientific Bias", in F. O. Schmitt (ed.), *The Neurosciences* (New York: Rockefeller University Press).

MACNAMARA, J. (1982), *Names for Things: A Study of Child Language* (Cambridge, Mass: MIT Press).

——(1986), *A Border Dispute: The Place of Logic in Psychology* (Cambridge, Mass.: MIT Press).

MACPHAIL, E. M. (1982), *Brain and Intelligence in Vertebrates* (Oxford: Clarendon Press).

——(1987), "The Comparative Psychology of Intelligence", *Behavioral and Brain Sciences*, 10: 645–95.

——(1994), *The Neuroscience of Animal Intelligence* (New York: Columbia University Press).

MANDLER, G. (1984), *Mind and Body: Psychology of Emotion and Stress* (New York: Norton).

MARLER, P. (1970), "Birdsong and Speech Development: Could there be Parallels?", *American Scientist*, 58: 669–73.

MARR, D. (1982), *Vision: A Computational Investigation into the Human Representation and Processing of Visual Information* (San Francisco: W. H. Freeman).

MATARIC, M. J. (1991), "Navigating with a Rat Brain: A Neurobiologically Inspired Model for Robot Spatial Representation", in J.-A. Meyer and S. W. Wilson (eds.),

From Animals to Animats: Proceedings of the First International Conference on Simulation of Adaptive Behavior (Cambridge, Mass.: MIT Press), 169–75.

MATIN, L., STEVENS, J. K., and PICOULT, E. (1983), "Perceptual Consequences of Experimental Extraocular Muscle Paralysis", in A. Hein and M. Jeannerod (eds.), *Spatially Oriented Behavior* (New York: Springer-Verlag), 243–62.

MATSUZAWA, T. (1985), "Use of Numbers by a Chimpanzee", *Nature*, 315: 57–9.

MATURANA, H. R., and VARELA, F. J. (1980), *Autopoiesis and Cognition: The Realization of the Living* (Boston: Reidel).

MAZZONI, P. and ANDERSEN, R. (1995), "Gaze Coding in the Posterior Parietal Cortex", in M. Arbib (ed.), *The Handbook of Brain Theory and Neural Networks* (Cambridge, Mass: MIT Press).

MECK, W. H., and CHURCH, R. M. (1983), "A Mode Control of Counting and Timing Processes", *Journal of Experimental Psychology: Animal Behavior Processes*, 9: 320–34.

MILLIKAN, R. G. (1984), *Language, Thought, and Other Biological Categories* (Cambridge, Mass.: MIT Press).

MILNER, A. D., and GOODALE, M. A. (1995), *The Visual Brain in Action* (Oxford: Oxford University Press).

MORRIS, M. R. (1992), *The Good and the True* (Oxford: Clarendon Press).

NATALE, F., and ANTINUCCI, F. (1989), "Stage 6 Object-Concept Representation", in F. Antinucci (ed.), *Cognitive Structure and Development in Nonhuman Primates* (Hillsdale, NJ: Lawrence Erlbaum Associates), 97–112.

NEWELL. A. (1980), "Physical Symbol Systems", *Cognitive Science*, 4: 135–83.

—— and SIMON, H. A. (1972), *Human Problem Solving* (Englewood Cliffs, NJ: Prentice-Hall).

NIGRO, G., and NEISSER, U. (1983), "Point of View in Personal Memories", *Cognitive Psychology*, 15: 467–82.

OLDS, J., and MILNER, P. (1954), "Positive Reinforcement Produced by Electrical Stimulation of Septal Area and Other Regions of Rat Brain", *Journal of Comparative Physiology and Psychology*, 47: 419–27.

ONO, M., KUBIK, S., and ABERNATHY, C. D. (1990), *Atlas of the Cerebral Sulci* (New York: Thieme Medical Publishers).

PACKARD, M., and TEATHER, L. (1998a), "Amygdala Modulation of Multiple Memory Systems: Hippocampus and Caudate-Putamen", *Neurobiology of Learning and Memory*, 69: 163–203.

—— —— (1998b), "Double Dissociation of Hippocampal and Dorsal-Striatal Memory Systems by Post-Training Intracerebral Injections of 2–amino-phosophono-pentanoic Acid", *Behavioral Neuroscience*, 111: 543–51.

PANKSEPP, J. (1991), "Affective Neuroscience", in T. K. Strongman (ed.), *International Review of Studies on Emotion* (New York: John Wiley).

PAPEZ, J. W. (1937), "A Proposed Mechanism of Emotion", *Archives of Neurology and Psychiatry*, 38: 725–44.

PASHLER, H. E. (1998), *The Psychology of Attention* (Cambridge, Mass.: MIT Press/Bradford Books).

PEACOCKE, C. A. B. (1997), "Metaphysical Necessity: Understanding, Truth and Epistemology", *Mind*, 106: 521–74.

—— (1999), *Being Known* (Oxford: Oxford University Press).

PEPPERBERG, I. M. (1994), "Numerical Competence in an African Gray Parrot (*Psittacus erithacus*), *Journal of Comparative Psychology*, 108: 36–44.

PERNER, J. (forthcoming), "Memory and Theory of Mind", in E. Tulving and F. I. M. Craik (eds.), *The Oxford Handbook of Memory* (Oxford: Oxford University Press).

——and Ruffman, T. (1995), "Episodic Memory and Autonoetic Consciousness: Developmental Evidence and a Theory of Childhood Amnesia", *Journal of Experimental Child Psychology*, 59: 516–48.

PERRY, R. B. (1921*a*), "A Behavioristic View of Purpose", *Journal of Philosophy*, 18: 85–105.

——(1921*b*), "The Cognitive Interest and its Refinement", *Journal of Philosophy*, 18: 365–75.

——(1921*c*), "The Independent Variability of Purpose and Belief", *Journal of Philosophy*, 18: 169–80.

PETERSEN, S. E., FOX, P. T., POSNER, M. I., MINTUN, M., and RAICHLE, M. E. (1988), "Positron Emission Tomographic Studies of the Cortical Anatomy of Single-Word Processing", *Nature*, 331: 585.

PIAGET, J. (1955), *The Child's Construction of Reality* (London: Routledge and Kegan Paul).

PINKER, S. (1994), *The Language Instinct* (New York: Morrow).

——(1997), *How the Mind Works* (New York: Norton).

POLLEN, D. (1995), "Cortical Areas in Visual Awareness", *Nature*, 377: 293–4.

POSNER, M. I., SNYDER, C., and DAVIDSON, B. (1980), "Attention and the Detection of Signals", *Journal of Experimental Psychology: General*, 109: 160–74.

——WALKER, J. A., FRIEDRICH, F. J., and RAFAL, R. D. (1984), "Effects of Parietal Injury on Covert Orienting of Visual Attention", *Journal of Neuroscience*, 4: 1863–74.

POUGET, A., and SEJNOWSKI, T. J. (1997*a*), "Lesion in a Basis Function Model of Parietal Cortex: Comparison with Hemineglect", in P. Their and H.-O. Karnath (eds.), *Parietal Lobe Contributions to Orientation in 3D Space* (Heidelberg: Springer-Verlag).

————(1997*b*), "A New View of Hemineglect Based on the Response Properties of Parietal Neurones", *Transactions of the Royal Society*, series B, 352: 1449–59.

————(1997*c*), "Spatial Transformations in the Parietal Cortex Using Basis Functions", *Journal of Cognitive Neuroscience*, 9/2: 222–37.

PREMACK, D. (1986), *Gavagai!* (Cambridge, Mass: MIT Press).

——and PREMACK, A. J. (1994*a*), "Origins of Human Social Competence", in M. Gazzaniga (ed.), *The Cognitive Neurosciences* (Cambridge, Mass.: MIT Press), 205–18.

————(1994*b*), "Levels of Causal Understanding in Chimpanzees and Children", *Cognition*, 50: 347–62.

PUTNAM, H. (1962), "Dreaming and 'Depth-Grammar'", in R. J. Butler (ed.), *Analytical Philosophy* (Oxford: Basil Blackwell), 211–35.

——(1978), *Meaning and the Moral Sciences* (London: Routledge and Kegan Paul).

PYLYSHYN, Z. W. (1984), *Computation and Cognition: Toward a Foundation for Cognitive Science* (Cambridge, Mass: MIT Press).

——(ed.) (1987), *The Robot's Dilemma: The Frame Problem in Artificial Intelligence* (Norwood, NJ: Ablex).

——(1989), "The Role of Location Indexes in Spatial Perception: A Sketch of the FINST Spatial Index Model", *Cognition*, 32: 65–97.

——(1998), "Visual Indexes in Spatial Attention and Imagery", in R. Wright (ed.), *Visual Attention* (Oxford: Oxford University Press), 187–214.

PYLYSHYN, Z. W. (1999), "Is Vision Continuous with Cognition? The Case for Cognitive Impenetrability of Visual Perception", *Behavioral and Brain Sciences*, 22: 341–423.

—— (forthcoming), "Seeing: It's not what you think", (Book MS).

—— and STORM, R. W. (1988), "Tracking of Multiple Independent Targets: Evidence for a Parallel Tracking Mechanism", *Spatial Vision*, 3: 1–19.

—— BURKELL, J., FISHER, B., SEARS, C., SCHMIDT, W., and TRICK, L. (1994), "Multiple Parallel Access in Visual Attention", *Canadian Journal of Experimental Psychology*, 48: 260–83.

QUINE, W. V. (1960), *Word and Object* (Cambridge, Mass.: MIT Press).

RAFFMAN, D. (1995), "On the Persistence of Phenomenology", in T. Metzinger (ed.), *Conscious Experience* (Schoningh).

RAMSEY, F. P. (1990), *Philosophical Papers*, ed. D. H. Mellor (Cambridge: Cambridge University Press).

RAPCSAK, S. Z., COMER, J. F., and RUBENS, A. B. (1993), "Anomia for Facial Expressions: Neuropsychological Mechanisms and Anatomical Correlates", *Brain and Language*, 45: 233–52.

RAY, T. S. (1990), "An Approach to the Synthesis of Life", in C. G. Langton, C. Taylor, J. D. F. and S. Rasmussen (eds.), *Artificial Life II* (Redwood City, Calif.: Addison-Wesley). Reprinted in M. A. Boden (ed.), *The Philosophy of Artificial Life* (Oxford: Oxford University Press, 1996), 111–45.

—— (1994), "An Evolutionary Approach to Synthetic Biology: Zen and the Art of Creating Life", *Artificial Life*, 1: 179–210.

RICHARDS, W. (1988), *Natural Computation* (Cambridge, Mass.: MIT Press).

RIDLEY, M. (1983), *The Explanation of Organic Diversity* (Oxford: Clarendon Press).

ROBERTSON, L., TREISMAN, A., FRIEDMAN-HILL, S., and GRABOWECKY, M. (1997), "The Interaction of Spatial and Object Pathways: Evidence from Balint's Syndrome", *Journal of Cognitive Neuroscience*, 9/3: 295–317.

ROCK, I., and GUTMAN, D. (1981), "The Effect of Inattention on Form Perception", *Journal of Experimental Psychology: Human Perception and Performance*, 7: 260–83.

ROITBLAT, H. L., HERMAN, M., and NACHTIGAL, P. E. (1993), *Language and Communication: Comparative Perspectives* (Hillsdale, NJ: Lawrence Erlbaum Associates).

ROLLS, E. T. (1992), "Neurophysiology and Functions of the Primate Amygdala", in J. P. Aggleton (ed.), *The Amygdala: Neurobiological Aspects of Emotion, Memory, and Mental Dysfunction* (New York: Wiley-Liss).

ROSENBERG, C. R., and SEJNOWSKI, T. J. (1990), "Parallel Networks that Learn to Pronounce English Text", *Complex Systems*, 1: 145–68.

ROSS, E. D., and MESULAM, M. M. (1979), "Dominant Language Functions of the Right Hemisphere? Prosody and Emotional Gesturing", *Archives of Neurology*, 36: 144–8.

RUMBAUGH, D. M., and WASHBURN, D. A. (1993), "Counting by Chimpanzees and Ordinality Judgements by Macaques in Video-formatted Tasks", in S. T. Boyseu and E. J. Capaldi (eds.), The Development of Numerical Competence: Animal and Human Models (Hillsdale, NJ: Lawrence Erlbaum Associates).

SAGI, D., and JULESZ, B. (1984), "Detection versus Discrimination of Visual Orientation", *Perception*, 13: 619–28.

SAVER, J. L., and DAMASIO, A. R. (1991), "Preserved Access and Processing of Social Knowledge in a Patient with Acquired Sociopathy Due to Ventromedial Frontal Damage", *Neuropsychologia*, 29: 1241–9.

SCHACTER, D. (1996), *Searching for Memory: The Brain, the Mind, and the Past* (New York: Basic Books).

SCHACHTER, S., and SINGER, J. (1962), "Cognitive, Social, and Physiological Determinants of Emotional State", *Psychological Review*, 69: 379–99.

SCHIFFER, S. (1987), *Remnants of Meaning* (Cambridge, Mass.: MIT Press/Bradford Books).

SCHOLL., B., and PYLYSHYN, Z. W. (1999), "Tracking Multiple Items through Occlusions: Clues to Visual Objecthood", *Cognitive Psychology*, 38: 259–90.

SEARLE, J. R. (1969), *Speech Acts, An Essay in the Philosophy of Language* (Cambridge: Cambridge University Press).

—— (1980a), "Minds, Brains, and Programs", *Behavioral and Brain Sciences*, 3: 417–24. Reprinted in M. A. Boden (ed.), *The Philosophy of Artificial Intelligence* (Oxford: Oxford University Press, 1990), 67–88.

—— (1980b), "Prima Facie Reasons", in Z. van Straaten (ed.), *Philosophical Subjects* (Oxford: Oxford University Press), 238–59.

—— (1990), "Who is Computing with the Brain?", *Behavioral and Brain Sciences*, 13/4: 632–4.

—— (1992), *The Rediscovery of the Mind* (Cambridge, Mass.: MIT Press).

—— (1995), *The Construction of Social Reality* (New York: Free Press).

SEARS, C. R., and PYLYSHYN, Z. W. (2000), "Multiple Object Tracking and Attentional Processes", *Canadian Journal of Experimental Psychology*, 54: 1–14.

SIMON, H. A. (1997), "Motivational and Emotional Controls of Cognition", in H. A. Simon, *Models of Thought* (New Haven: Yale University Press), 23–8.

SIMON, T. J. (1997), "Reconceptualizing the Origins of Number Knowledge: A 'Nonnumerical' Account", *Cognitive Development*, 12: 349–72.

—— and VAISHNAVI, S. (1996), "Subitizing and Counting Depend on Different Attentional Mechanisms: Evidence from Visual Enumeration in Afterimages", *Perception and Psychophysics*, 58/6: 915–26.

—— HESPOS, S., and ROCHAT, P. (1995), "Do Infants Understand Simple Arithmetic? A Replication of Wynn 1992", *Cognitive Development*, 10: 253–69.

SIMS, K. (1994), "Evolving 3D Morphology and Behavior by Competition", *Artificial Life*, 1: 353–72.

SNODGRASS, J. G., and VANDERWART, M. (1980), "A Standardized Set of 260 Pictures: Norms for Name Agreement, Image Agreement, Familiarity and Visual Complexity", *Journal of Experimental Psychology: Human Learning and Memory*, 6: 174–215.

SPELKE, E. S. (1985), "Preferential Looking Methods as Tools for the Study of Cognition in Infancy", in G. Gottlieb and N. Krasnegor (eds.), *Measurement of Audition and Vision in the First Year of Post-Natal Life* (Hillsdale, NJ: Lawrence Erlbaum Associates), 323–64.

—— (1991), "Physical Knowledge in Infancy: Reflections on Piaget's Theory", in S. Carey and R. Gelman (eds.), *The Epigenesis of Mind: Essays on Biology and Cognition* (Hillsdale, NJ: Lawrence Erlbaum Associates), 37–61.

—— (1994), "Initial Knowledge: Six Suggestions", *Cognition*, 50: 431–45.

—— VISHTON, P., and VON HOFSTEN, C. (1995), "Object Perception, Object-Directed Action, and Physical Knowledge in Infancy", in M. Gazzaniga (ed.), *The Cognitive Neurosciences* (Cambridge, Mass.: MIT Press), 165–79.

—— BREINLINGER, H., MACOMBER, J., and JACOBSON, K. (1992), "Origins of Knowledge", *Psychological Review*, 99: 603–32.

—— Kestenbaum, R., Simons, D. J., and Wein, D. (1995), "Spatio-Temporal Continuity, Smoothness of Motion and Object Identity in Infancy", *British Journal of Developmental Psychology*, 13: 113–42.

Sperry, R. W. (1981), "Cerebral Organization and Behavior", *Science*, 133: 1749–57.

Sperry, R. W., Gazzaniga, M. S., and Bogen, J. E. (1969), "Interhemispheric Relationships: The Neocortical Commissures; syndromes of their disconnection", in P. J. Vinken and G. W. Bruyn (eds.), *Handbook of Clinical Neurology*, iv (Amsterdam: North Holland, 273–90.

—— Zaidel, E., and Zaidel, D. (1979), "Self Recognition and Social Awareness in the Deconnected Minor Hemisphere", *Neuropsychologia*, 17: 153–66.

Stalnaker, R. (1998), "On the Representation of Context", *Journal of Logic, Language and Information*, 7/1: 3–16.

Starkey, P. (1992), "The Early Development of Numerical Reasoning", *Cognition*, 43: 93–126.

—— and Cooper, R. (1980), "Perception of Numbers by Human Infants", *Science*, 210: 1033–5.

Stiles, J., and Thal, D. (1993), "Linguistic and Spatial Cognitive Development Following Early Focal Brain Injury: Patterns of Deficit and Recovery", in M. Johnson (ed.), *Brain Development and Cognition: A Reader* (Oxford: Basil Blackwell).

Strauss, M., and Curtis, R. (1981), "Infant Perception of Numerosity", *Child Development*, 52: 1146–53.

Talairach, J., and Tournoux, P. (1988), *Co-planar Stereotaxic Atlas of the Human Brain. 3-Dimensional Proportional System: An Approach to Cerebral Imaging* (New York: Thieme Medical Publishers).

Taylor, C. (1971), "Interpretation and the Sciences of Man", *Review of Metaphysics*, 25: 3–51.

Thomas, R. K. (1992), "Conceptual Use of Number: Ecological Perspectives and Psychological Processes", in T. Nishida, W. C. McGrew, P. Marler, M. Pickford, and F. B. M. de Waal (eds.), *Topics in Primatology*, i: *Human Origins* (Tokyo: Tokyo University Press), 305–14.

Thompson, R. K. R. (1995), "Natural and Relational Concepts in Animals", in H. L. Roitblat and J. A. Meyer (eds.), *Comparative Approaches to Cognitive Science* (Cambridge, Mass.: MIT Press), 175–224.

Tinkelpaugh, O. (1928), "An Experimental Study of Representative Factors in Monkeys", *Journal of Comparative Psychology*, 8: 197–236.

—— (1932), "Multiple Delayed Reaction with Chimpanzees and Monkeys", *Journal of Comparative Psychology*, 13: 207–43.

Tipper, S. P., and Behremann, M. (1996), "Object- Centered not Scene-Based Visual Neglect", *Journal of Experimental Psychology: Human Perception and Performance*, 22/5: 1261–78.

Tolman, E. C. (1922), "A New Formula for Behaviorism", *Psychological Review*, 29: 44–53.

Tranel, D. (1996), "The Iowa–Benton School of Neuropsychological Assessment", in I. Grant and K. M. Adams (eds.), *Neuropsychological Assessment of Neuropsychiatric Disorders*, 2nd edn. (New York: Oxford University Press), 81–101.

—— Damasio, H., and Damasio, A. R. (1995), "Double Dissociation between Overt and Covert Face Recognition", *Journal of Cognitive Neuroscience*, 7: 425–32.

—— —— —— (1997), "A Neural Basis for the Retrieval of Conceptual Knowledge", *Neuropsychologia*, 35: 1319–27.

——LOGAN, C. G., FRANK, R. J., and DAMASIO, A. R. (1997), "Explaining Categ-ory-Related Effects in the Retrieval of Conceptual and Lexical Knowledge for Concrete Entities: Operationalization and Analysis of Factors", *Neuropsychologia*, 35: 1329–39.

TREISMAN, A. (1988), "Features and Objects: The Fourteenth Bartlett Memorial Lecture", *Quarterly Journal of Experimental Psychology*, 40A: 201–37.

——and CELADE, G. (1980), "A Feature Integration Theory of Attention", *Cognitive Psychology*, 12: 97–136.

——and KAHNEMAN, D. (1992), "The Reviewing of Object Files: Object-Specific Integration of Information", *Cognitive Psychology*, 24: 175–219.

——and SCHMIDT, H. (1982), "Illusory Conjunctions in the Perception of Objects", *Cognitive Psychology*, 14: 107–14.

——and SOUTHER, H. (1986), "Illusory Words: The Roles of Attention and of Top–Down Constraints in Conjoining Letters to Form Words", *Journal of Experimental Psychology: Human Perception and Performance*, 12: 3–17.

TRICK, L., and PYLYSHYN, Z. W. (1993), "What Enumeration Studies Tell us about Spatial Attention: Evidence for Limited Capacity Preattentive Processing", *Journal of Experimental Psychology: Human Perception and Performance*, 19: 331–51.

————(1994*a*), "Cuing and Counting: Does the Position of the Attentional Focus Affect Enumeration?", *Visual Cognition*, 1: 67–100.

————(1994*b*), "Why are Small and Large Numbers Enumerated Differently? A Limited Capacity Preattentive Stage in Vision", *Psychological Review*, 10: 1–23.

ULLER, C., HAUSER, M., and CAREY, S. (forthcoming), "The Spontaneous Represen-tation of Number in a New World Primate Species, Cotton Top Tamarins", *Journal of Comparative Psychology*.

——HUNTLEY-FENNER, G., CAREY, S., and KLATT, L. (1999), "What Representations Might Underlie Infant Numerical Knowledge?", *Cognitive Development*, 14: 1–36.

——XU, F., CAREY, S., and HAUSER, M. (1997), "Is Language Needed for Construct-ing Sortal Concepts?" A Study with Nonhuman Primates", *Proceedings of the 21st Annual Boston University Conference on Language Development*, ii (Somerville, Mass.: Cascadilla Press), 665–7.

ULLMAN, S. (1984), "Visual Routines", *Cognition*, 18: 97–159.

UNGERLEIDER, L. G., and MISHKIN, M. (1982), "Two Cortical Visual Systems", in J. Ingle, M. A. Goodale, and R. J. W. Mansfield (eds.), *Analysis of Visual Behavior* (Cambridge, Mass.: MIT Press), 549–86.

VAN LOOSBROEK, E., and SMITSMAN, A. (1990), "Visual Perception of Numerosity", *Developmental Psychology*, 26: 916–22.

WATSON, D. G., and HUMPHREYS, G. W. (1997), "Visual Marking: Prioritizing Selec-tion for New Objects by Top–Down Attentional Inhibition of Old Objects", *Psy-chological Review*, 104: 90–122.

WERNICKE, C. (1874), *Der aphasische Symptomencomplex* (Breslau: Cohn and Weigert).

WIGGINS, D. (1980), *Sameness and Substance* (Oxford: Basil Blackwell).

WILSON, M. A., and MCNAUGHTEN, B. L. (1993), "Dynamics of the Hippocampal Ensemble Code for Space", *Science*, 261: 1055–8.

————(1994), "Reactivation of Hippocampal Ensemble Memories during Sleep", *Science*, 265: 676–9.

WIMMER, H., HOGREFE, G.-J., and PERNER, J. (1988), "Children's Understanding of Information Access as Source of Knowledge", *Child Development*, 59: 386–96.

WIMMER, H., HOGREFE, G.-J., and SODIAN, B. (1988), "A Second Stage in Children's Conception of Mental Life: Understanding Sources of Information", in J. Astington, P. Harris, and D. Olson (eds.), *Developing Theories of Mind* (New York: Cambridge University Press).

WISE, S. P., BOUSSAOUD, D., JOHNSON, P. B., and CAMINITI, R. (1997), "Premotor and Parietal Cortex: Corticocortical Connectivity and Combinatorial Computations", in W. Cowan, E. Shooter, C. Stevens and R. Thompson (eds.), *Annual Review of Neuroscience*, 20: 25–42.

WISER, M., and CAREY, S. (1983), "When Heat and Temperature Were One", in D. Gentner and A. Stevens (eds.), *Mental Models* (Hillsdale, NJ: Lawrence Erlbaum Associates).

WOODS, R. P., MAZZIOTTA, J. C., and CHERRY, S. R. (1993), "MRI-PET Registration with Automated Algorithm", *Journal of Computer Assisted Tomography*, 17: 536–46.

WORSELEY, K. J. (1994), "Local Maxima and the Expected Euler Characteristic of Excursion Sets of c^2 F, and t fields", *Adv. Appl. Probl.* 26: 13–42.

——EVANS, A. C., MARRETT, S., and NEELIN, P. (1992), "A Three-Dimensional Statistical Analysis for CBF Activation Studies in Human Brain", *Journal of Cerebral Blood Flow Metabolism*, 12: 900–18.

WYNN, K. (1992), "Addition and Subtraction by Human Infants", *Nature*, 358: 749–50.

——(1996), "Infant's Individuation and Enumeration of Actions", *Psychological Science*, 7: 164–9.

XU, F. (1997), "From Lot's Wife to a Pillar of Salt: Evidence that *Physical Object* is a Sortal Concept", *Mind and Language*, 12: 365–92.

——and CAREY, S. (1996), "Infants' Metaphysics: The Case of Numerical Identity", *Cognitive Psychology*, 30: 111–53.

——CAREY, S., and WELCH, J. (1999), "Infants' Ability to Use Object Kind Information for Object Individuation", *Cognition*, 70: 137–66.

YANTIS, S. (1992), "Multielement Visual Tracking: Attention and Perceptual Organization", *Cognitive Psychology*, 24: 295–340.

——(1998), "Objects, Attention and Perceptual Experience", in R. Wright (ed.), *Visual Attention*, (Oxford: Oxford University Press), 187–214.

——and GIBSON, B. S. (1994), "Object Continuity in Apparent Motion and Attention", *Canadian Journal of Experimental Psychology*, 48/2: 182–204.

——and JOHNSON, D. N. (1990), "Mechanisms of Attentional Priority", *Journal of Experimental Psychology: Human Perception and Performance*, 16/4: 812–25.

——and JONES, E. (1991), "Mechanisms of Attentional Selection: Temporally Modulated Priority Tags", *Perception and Psychophysics*, 50: 166–78.

ZAJONC, R. B. (1984), "On the Primacy of Affect", *American Psychologist*, 39: 117–23.

ZHANG, K.-C. (1996), "Representation of Spatial Orientation by the Intrinsic Dynamics of the Head-Direction Cell Ensemble: A Theory", *Journal of Neuroscience*, 16: 2112–26.

——GINZBURG, I., McNAUGHTEN, B. L., and SEJNOWSKI, T. (1998), "Interpreting Neuronal Population Activity by Reconstruction: A Unified Framework with Application to Hippocampal Place Cells", *Journal of Neurophysiology*.

ZIPSER, D., and ANDERSEN, R. A. (1988), "A Back-Propagation Network that Simulates Response Properties of a Subset of Posterior Parietal Neurons", *Nature*, 331: 679–84.

Index